Gabriel Wurzer • Kerstin Kowarık
Hans Reschreiter

Editors

Agent-based Modeling and Simulation in Archaeology

 Springer

Editors
Gabriel Wurzer
Digital Architecture and Planning
Institute of Architectural
  Sciences
Vienna, Austria

Kerstin Kowarik
Prehistory
Natural History Museum Vienna
Vienna, Austria

Hans Reschreiter
Prehistory
Natural History Museum Vienna
Vienna, Austria

ISSN 1867-2434   ISSN 1867-2442 (electronic)
ISBN 978-3-319-00007-7   ISBN 978-3-319-00008-4 (eBook)
DOI 10.1007/978-3-319-00008-4
Springer Cham Heidelberg New York Dordrecht London

Library of Congress Control Number: 2014951909

Printed on acid-free paper

Springer is part of Springer Science+Business Media (www.springer.com)

Preface

In March 2011, we held a workshop called "Agents in Archaeology" at the Natural History Museum Vienna, in which we brought together both experts and novices in archaeological simulation: On the one hand, we had a range of presentations on the practical use of agent-based modelling *as a research tool*; on the other hand, we held a 2-day tutorial on NetLogo for archaeologists not familiar with simulation software (or even programming). As supplement, we also uploaded the tutorial lecture to YouTube, which has so far attracted 6,206 viewers devoting 35,629 minutes of their lifetime to that subject. Hence, when Suzana Dragicevic invited us on behalf of the Springer GIScience series to compile a book on that matter, we knew that we would like to keep this very delicate balance between practical "hands-on"-type contributions (given by "digital archaeologists", if you will) and methodical chapters (given by modelling experts, mathematicians, computer scientists and social scientists). Accordingly, we decomposed the subject into four parts: *Introduction* (which gives an overview), *Methods* (which elaborates the foundations of the subject), *Applications* (which reports on real models, in a "hands-on" fashion) and *Summary and Outlook* (which gives some future trends).

Now that the volume is finished, it is time to take a step back and look at the results. Our authors have given an excellent view *behind the curtain*, not only technically/methodically but also concerning the *development process of a model* (e.g. the Artificial Anasazi in Chap. 2 or the Sugarscape model in Chap. 11). They did so using 733 citations which we have plotted in Fig. 1: Each dot corresponds to one citation; dots of larger size are those citations that are referenced by multiple chapters and are thus deemed as being *highly influential work* for this book. In more detail, we can see a cluster of work around the year 2000 (e.g. Sugarscape and NetLogo, both having five citations) and also one at 2007 (e.g. a postpositivist view on agent-based modelling). For us, the clusters at the turn of the millennium concerns pioneering work that is still highly influential, whereas the second cluster is a body of work that goes into the direction of a differentiation and specialisation of modelling. Interestingly, the number of cited publications is rising each year, but the amount of highly influential publications does not rise with it (it rather seems to be constant). Surely, there is also a time factor in this (more recent publications are not cited as often), but could it possibly mean that the field is returning to models it has gotten used to, only on a broader basis? Or, put differently: Are some types of models evolving into a quasi-standard? What is the modelling philosophy that we buy ourselves into, if that is true? Partial answers to these questions appear in the subsequent chapters (especially Chaps. 1 and 11); however, these entail new questions leading to new lines of thought.

In this sense, we hope that we can contribute one next step to the field of digital archaeology with our book. We would like to warmly thank Suzana Dragicevic and Ron Doering for making this work possible and our colleagues and families for supporting us. Last but not least, we thank all contributing authors for their excellent work.

Vienna, Austria   Gabriel Wurzer
Vienna, Austria   Kerstin Kowarik
Vienna, Austria   Hans Reschreiter
January 2014

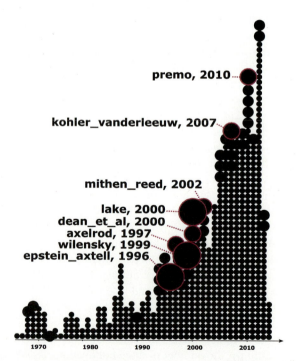

Fig. 1 Analysis of all 733 citations in the book, shown yearly from 1970 onwards: one *dot* per citation; *dot size* reflects the number of chapters referencing the citation

Contents

Part IV   Summary and Outlook

Contributors

Joan A. Barceló Universitat Autònoma de Barcelona, Barcelona, Spain

Martin Bicher Vienna University of Technology, Vienna, Austria

Felix Breitenecker Vienna University of Technology, Vienna, Austria

Richard Cimler University of Hradec Králové, Hradec Králové, Czech Republic

André Costopoulos McGill University, Montréal, QC, Canada

Enrico R. Crema University College London, London, UK

Alžběta Danielisová Institute of Archaeology of the Czech Academy of Sciences, Prague, Czech Republic

Florencia Del Castillo Universitat Autìnoma de Barcelona, Barcelona, Spain

Ricardo Del Olmo Universidad de Burgos, Burgos, Spain

George J. Gumerman Santa Fe Institute, Santa Fe, NM, USA

Andreas Koch University of Salzburg, Salzburg, Austria

Kerstin Kowarik Prehistory, Natural History Museum Vienna, Vienna, Austria

Mark W. Lake UCL Institute of Archaeology, London, UK

Tomáš Machálek University of Hradec Králové, Hradec Králové, Czech Republic

Laura Mameli Universitat de Girona, Girona, Spain

Francisco J. Miguel Quesada Universidad Autìnoma de Barcelona, Barcelona, Spain

Kamila Olševičová University of Hradec Králové, Hradec Králové, Czech Republic

Philipp Pichler DWH Simulation Services, Vienna, Austria

Niki Popper DWH Simulation Services, Vienna, Austria

David Poza Universidad de Valladolid, Valladolid, Spain

Hans Reschreiter Prehistory, Natural History Museum Vienna, Vienna, Austria

Xavier Rubio-Campillo Barcelona Supercomputing Center, Barcelona, Spain

Lisa Sattenspiel University of Missouri-Columbia, Columbia, MO, USA

Alan C. Swedlund University of Massachusetts-Amherst, Amherst, MA, USA

Xavier Vilà Universidad Autnoma de Barcelona, Barcelona, Spain

Amy L. Warren University of Missouri-Columbia, Columbia, MO, USA

Gabriel Wurzer Vienna University of Technology, Vienna, Austria

Part I
Introduction

Chapter 1
Explaining the Past with ABM: On Modelling Philosophy

Mark W. Lake

1.1 Introduction

This chapter discusses some of the conceptual issues surrounding the use of agent-based modelling in archaeology. Specifically, it addresses three questions: Why use agent-based simulation? Does specifically *agent-based* simulation imply a particular view of the world? How do we learn by simulating? First, however, it will be useful to provide a brief introduction to agent-based simulation and how it relates to archaeological simulation more generally. Some readers may prefer to return to this chapter after having read a more detailed account of an exemplar (Chap. 2) or of the technology (Chap. 3). Textbooks on agent-based modelling include Grimm and Railsback [(2005) Individual-based modeling and ecology, Princeton University Press, Princeton] and Railsback and Grimm [(2012) Agent-based and individual-based modeling: a practical introduction, Princeton University Press, Princeton], both aimed at ecologists, the rather briefer [Gilbert (2008) Agent-based models. Quantitative applications in the social sciences, Sage, Thousand Oaks, CA], aimed at sociologists, and [Ferber (1999) Multi-agent systems: an introduction to distributed artificial intelligence, English edn. Addison-Wesley, Harlow], which treats agent-based simulation from the perspective of artificial intelligence and computer science.

M.W. Lake (✉)
UCL Institute of Archaeology, Gordon Square, London, UK
e-mail: mark.lake@ucl.ac.uk

© Springer International Publishing Switzerland 2015
G. Wurzer et al. (eds.), *Agent-based Modeling and Simulation in Archaeology*,
Advances in Geographic Information Science, DOI 10.1007/978-3-319-00008-4_1

1.2   Agent-Based Modelling in Archaeology

Agent-based modelling (ABM) is a method of computer simulation that is particularly well suited to exploring how the aggregate characteristics of some system arise from the behaviour of its parts. The parts in question are modelled as 'agents', that is, units which have causal efficacy and can reasonably be treated as individuals in the sense that they act as cohesive wholes in respect of the particular research problem. Agents are usually situated in an artificial environment and their behaviour is governed by rules which specify how they respond to the content of that environment and possibly also the behaviour of other agents (Epstein and Axtell 1996, p. 5).

The well known Long House Valley agent-based model (chapter 2; Dean et al. 2000; Axtell et al. 2002; also Kohler et al. 2005 for a popular account) illustrates the paradigmatic features of a typical agent-based model. The Long House Valley ABM was built to explore the relationship between climatically determined resource availability, settlement location and population growth in Long House Valley, Arizona in the period AD 400–1450. In this example the agents are individual Puebloan households, each of which use rules to choose where to settle in Long House Valley in order to grow sufficient maize to survive. The maize-growing potential for each hectare in the valley was inferred from painstaking palaeoenvironmental research and is input into the model for every year from AD 400 to 1450. When the simulation is run, growth of a household and/or environmental degradation can lead to fissioning or wholesale relocation of that settlement. In this way, repeated individual household decision-making and reproduction produces a changing macro-level settlement pattern and population size, both of which could be compared with the archaeological evidence. After experimenting with different assumptions concerning the productivity of prehistoric maize and the fertility and longevity of households, the authors concluded that climate change alone is insufficient to explain the eventual abandonment of Long House Valley.

The Long House Valley agents fulfil the standard textbook requirement (e.g. Ferber 1999, pp. 9–10) that agents should be autonomous (directed by their own goals), goal-directed (behave in an attempt to achieve their goals), reactive (change their behaviour in response to the properties of the environment) and—usually—situated (have an explicit location in the environment). It is, however, possible to add additional complexity designed to make agents more human-like. Thus Ferber (1999) notes that agents may additionally maintain a representation of their world and thus be capable of cognition, be social in the sense of interacting with and in particular maintaining patterned interaction with other agents, and be capable of reproduction involving some kind of recombination or mutation of their attributes. All of these 'extensions' have in fact been implemented in archaeological agent-based models, in some cases more than 20 years ago. Thus, for example, Mithen (1990) and Lake's (2000) agent-based models of Mesolithic hunter-gatherer foraging very explicitly model agents learning about their environment and using the information so gained to inform their decision-making; in the case of Lake's model this extends to each agent having a geographically referenced cognitive map of its environment. Both these models rely on agents having memory, but in another case the capacity and longevity of agent memory was itself the subject of enquiry (Costopoulos 2001). Similarly, a number of archaeological agent-based models have incorporated aspects of social interaction, ranging from inter-agent social learning (Kohler et al. 2012b; Lake 2000; Mithen 1989; Premo 2012; Premo and Scholnick 2011) through simple collective decision-making (Lake 2000) to exchange (Bentley et al. 2005; Kobti 2012), group formation (Doran et al. 1994; Doran and Palmer 1995) and the emergence of leaders (Kohler et al. 2012b). Agent reproduction involving recombination or mutation of attributes can also be found in archaeological agent-based models, including Kachel et al's (2011) evaluation of the so-called 'grandmother hypothesis' for human evolution, Lake's (2001b) model of the evolution of the hominin capacity for cultural learning and Xue et al's (2011) model of the extent to which achieving a high degree of adaption to the environment can be detrimental in the long term.

As well as making agents more human-like, it is also possible to extend a paradigmatic agent-based model by explicitly modelling feedback between agent behaviour and the properties of the environment in which the agents are situated. In a very simple agent-based model the environment might be completely unchanging. This was not the case in the Long House Valley model, since that model explicitly included external environmental forcing by altering the maize yield over time in accordance with the palaeoclimatic reconstruction. Other archaeological agent-based models make the environment even more dynamic by, for example, explicitly modelling resource degradation as a result of agent activity. One of the first models to achieve this was the first (mid 1990s) version of Kohler et al's Village ABM, which explicitly incorporated reduced yields from continued farming (2000; 2012a). More recent versions of the Village ABM also explicitly model the population growth of prey species such as deer (Johnson and Kohler 2012). Unsurprisingly, some of the most sophisticated modelling of environmental change is to be found in agent-based models conceived as part of the growing programme of 'socionatural' studies (McGlade 1995; van der Leeuw and Redman 2002; Kohler and van der Leeuw 2007a) of the *socioecological* (Barton et al. 2010a) dynamics of long-term human environment interaction. At present there is a concerted effort to further this agenda by coupling agent-based models of human behaviour with established geographical information systems or other raster models of natural processes such as soil erosion (e.g. Barton et al. 2010b,a; Kolm and Smith 2012).

I have recently provided an up-to-date review of the use of computer simulation in archaeology (Lake 2014), which includes detailed discussion of the vicissitudes and subject matter of archaeological agent-based modelling. For the purposes of this chapter it is worth drawing out three main points. The first is that archaeologists have used simulation models that exhibit many of the characteristics of agent-based models for almost as long as they have used computers, and certainly long before adoption of the term 'agent-based model' in the 1990s. Wobst considered that the approach taken in his 1974 computer simulation model of Palaeolithic social systems could "best be conceptualized as an educational game consisting of a gaming table (area), pieces (people), rules (biological or cultural rules of behavior), and a series of different outcomes depending on the specifications of the components" (Wobst 1974, p. 158); clearly, we would now call this an agent-based model, as Wobst has himself noted (2010, p. 9). Thomas' (1972) model of Shoshonean subsistence also exhibits some of the characteristics of an agent-based model (particularly in its implementation of a spatially heterogeneous environment) and Wright and Zeder's 1977 simulation model of linear exchange is another that could be considered to be an early agent-based model (Doran 2000, p. 90). Mithen's late 1980s simulation model of Mesolithic hunter-gather decision-making (Mithen 1987, 1990) is perhaps the most recent well-known agent-based simulation model that was not explicitly labelled as such. Even recent textbooks on ABM now acknowledge that "agent-based modelling is no longer a completely new approach" (Railsback and Grimm 2012, p. 11), but they also note the potential advantage that "the worst mistakes have been made and corrected" (ibid.). This, of course, will only be true if new modellers drawn in by ever more user-friendly software make the effort to acquaint themselves with earlier agent-based models and avoid naively re-inventing the wheel.

The second point that can be drawn from my review of archaeological simulation is that, although something like agent-based modelling has been used in archaeology for 40 years, there has unquestionably been an explosion of interest in the technique since the turn of the millennium. Kohler and Gumerman's (2000) influential collection of agent-based models, *Dynamics in Human and Primate Societies*, heralded the arrival in archaeology of fully modern and self-identified agent-based simulation. The influence of modern agent-based modelling on the resurgence of archaeological interest in computer simulation more generally can not be overstated: over 50 % of the *c.* 70 substantive archaeological simulation models published between 2001 and March 2013 are agent-based models (Lake 2014). These models differ widely in complexity, ranging from those that barely meet the minimum textbook definition of an agent-based model, but were implemented using the computational framework provided by an agent-based simulation toolkit (e.g. Bentley et al. 2004), through relatively simple abstract models that are however unquestionably agent-based (e.g. Premo 2007), to much more complex realistic models that exhibit many of the optional attributes discussed above (e.g. Kohler et al. 2012a; Wilkinson et al. 2007).

A final point to make about the status of agent-based modelling in archaeology is that it can now be argued to have acquired a degree of methodological maturity (Lake 2014). First, a high proportion of recent publications reporting simulation results do not foreground the method itself, or at the very least they balance the account of method with a substantive research conclusion. The latter point provides the second line of evidence that simulation has finally come of age as part of the archaeological toolkit: that archaeological simulation models increasingly provide results that are useful to researchers who were not involved in the modelling process (i.e. they have what Innis (1972, p. 34) referred to as "output utility"). That said, it is also clear that agent-based modelling is by no means evenly distributed across all areas of archaeological enquiry, but is largely concentrated in the study of human evolution, evolutionary archaeology and the aforementioned area of socionatural studies. Furthermore, these areas are in turn differentiated by the extent to which they favour relatively simple models offering a high level of abstraction (typical of human evolution and evolutionary archaeology) or more complex models offering greater realism (typical of socionatural studies), an issue to which I will return.

1.3 Why Agent-Based Modelling?

As just discussed, there has been a resurgence of interest in the application of computer simulation to archaeological problems, and agent-based models account for over half of all archaeological computer simulation undertaken in the new millennium. Even if this burgeoning activity reflects increased acceptance of the general case for using computer simulation—of whatever kind—in archaeology, it is very likely to have been triggered by the conceptual accessibility of agent-based modelling and/or the sense that agent-based modelling is part of a scientific paradigm that aligns well with contemporary archaeological interests. Each of these reasons for the growth of agent-based modelling is discussed in turn.

1.3.1 The General Case for Computer Simulation in Archaeology

The case for using computer simulation in archaeology has been well-rehearsed (e.g. initially in Doran (1970) and more recently in Kohler (2000) and Premo et al. (2005)) and has four main strands: enforcing conceptual clarity in the interest of 'doing science', helping understand how things change, helping infer past behaviour from a static archaeological record, and testing other quantitative methods.

The case for computer simulation starts with the observation that archaeologists routinely build models, that is "pieces of machinery that relate observations to theoretical ideas" (Clarke 1972, p. 2). As Kohler and van der Leeuw (2007a, p. 3) remind us, even informal explanations for "how" or "why" something happened in the past are in fact models. Of course, some models are more formal than others, although even formal models vary greatly in their means of expression, ranging from, for example, the material replica of an Iron Age roundhouse through algorithmic specification of hunter-gatherer decision-making to the use of coupled differential equations to study the rise of urbanism. The advantage of formal modelling is that, by making explicit and unambiguous the relationships between the things in the model and also the intended scope of the model, it is easier to determine whether the model is supposed to be applicable to some observed phenomenon and, if so, whether it adequately predicts or fits it. This in turn facilitates the pursuit of archaeology as a science, whether one wishes to test a hypothesis in accordance with the hypothetico-deductive framework of the New Archaeology (Watson et al. 1971), or explore the utility of a model in a manner more consistent with contemporary model-based science (see Kohler and van der Leeuw (2007a, p. 3) for a manifesto; also commentary in Sect. 1.5). Thus one of the most important benefits of computer simulation is simply that in order to implement a model as a computer programme it must be very precisely specified in mathematical or algorithmic terms (Doran 1970, p. 298). If the modeller learns something from this process then the simulation can be said to have "conceptual utility" (Innis 1972, p. 33) because it has served to "create new problems and view old ones in new and interesting ways" (Zubrow 1981, p. 143). Of course, actually solving problems requires an appropriate inferential strategy (Premo 2007), as discussed in Sect. 1.5.1.

While computer simulation models may be among the most formal and least ambiguous of all kinds of model, the characteristic which sets them apart from other kinds of models is, of course, that they are iterative (Clarke 1972, p. 2). One should, however, distinguish between iterative models in general (numerical models) and simulation models in particular (see Lake (2001a, p. 723–4) for more detailed discussion). Numerical modelling is often used to obtain an approximate solution to some analytically intractable mathematical model which has been designed to predict the equilibrium state of a system, but without any interest in how it came about (for example an optimal subsistence strategy, as per e.g. Belovsky 1987). In contrast, simulation models are explicitly concerned with the passage of time and the state of the system changes by a process that is in some way analogous to the process of change in the real world (Doran and Hodson (1975, p. 286) and Renfrew (1981, pp. 292–3)). This explicit modelling of *process* is important for two reasons.

First, computer simulation subtly shifts the focus of modelling from asking "how it works" to "how it got to be as it [was]" (Allen and McGlade 1987, p. 724). In this, simulation better aligns formal modelling with modern under-standing of the importance of non-linearity, recursion and noise in the evolution of living systems, whether that is couched in the language of chaos (Schuster 1988), complexity (Waldrop 1992), niche construction (Odling-Smee et al. 2003), structuration (Giddens 1984), evolutionary drive (Allen and McGlade 1987) or contingency (Gould 1989). Thus, to pick just three examples of *explicit* concern with past *dynamics*, archaeologists have built non-linear dynamical systems models to investigate the likelihood of *instability* in European Bronze Age exchange systems (McGlade 1997), and agent-based models to explore *cyclical* nucleation and dispersion in Jomon settlement (Crema 2013) and the optimal adaptive fit to environments punctuated by *rapid reversals* (Xue et al. 2011).

Second, and related, the focus of simulation on process fits well with the notion that archaeology has as much or more to offer contemporary society as a science of long-term societal change and human-environment interaction than it does as a provider of ethnographic-scale snapshots of the past. While the latter can usefully challenge us to recognise the diversity of human lifeways, it is only by studying *processes* that one can explain why particular lifeways obtained in particular circumstances. Furthermore, understanding the processes that drive social change and environmental adaptation affords the possibility of predicting how contemporary lifeways might evolve in the future, especially if one studies such processes in the spirit that "history is still running" (Allen et al. 2006, p. 2). Archaeologists working under the banner of socionatural studies are particularly conscious of this last point and have forcefully made the case for "an enhanced role for archaeologists in the study of contemporary environmental issues" (van der Leeuw and Redman 2002, abstract), even going so far as to suggest that modern archaeology has the potential to be "at the center of modern studies of long-term global change" (van der Leeuw 2008, p. 477). The intellectual antecedents of this agenda can be found in Leslie White's interest in the long-term evolution of energy capture and Julian Steward's awareness that environmental adaptation is mediated through material culture and knowledge (Kohler and van der Leeuw 2007a, p. 11), but the contemporary approach is augmented with more sophisticated understanding of the nature of change in complex systems, one which rejects the a priori assumption of a unilineal trajectory and which can now be better pursued through advances in computer simulation (see Barton (2013) for a recent manifesto and Kohler and Varien (2012) for the history and role of simulation in one long-running socionatural study).

While computer simulation helps archaeologists study how processes of social and environmental change unfold, and thus contributes to archaeology as a human science uniquely equipped to study long-term change in socionatural systems, it also contributes to the more specifically archaeological problem of inferring what specific processes, that is, behaviour, produced the observed archaeological evidence. Binford (1981) long ago made the argument that archaeologists must infer past dynamics (behaviours) from a static archaeological record and, of course, the usual means of doing this is to compare patterns in the archaeological evidence with those expected from the candidate behaviour. As Kohler et al. (2012a, p. 40) remind us, archaeologists usually make the connection between a particular behaviour and an expected archaeological pattern on the basis of "intuition or common sense, ethnographic analogies and environmental regularities, or in some cases experimental archaeology", but computer simulation offers another—sometimes the only—way of doing this. For example, Mithen's (1988; 1990) computer simulation of Mesolithic hunting goals generated a virtual faunal assemblage whose species composition could then be compared with the archaeological record. In similar vein, Premo's (2005) simulation of Pleistocene hominin food sharing created virtual assemblages which revealed that the dense artifact accumulations at Olduvai and Koobi Fora, long attributed to central place foraging, could alternatively have been formed by routed foraging in a patchy environment. These two examples demonstrate the advantage of adding computer simulation to the archaeologist's inferential toolkit: not only that it forces us to codify and make explicit our assumptions, but that it also allows us to explore the outcome of behaviours which can no longer be observed and for which there is no reliable recent historical record. In addition, it allows us to explore the outcome of behaviour aggregated at the often coarse grained spatial and temporal resolution of the archaeological record.

The last point, that the archaeological record provides only a—often sparse— sample of the outcome of whatever behaviour generated it, leads to another well established archaeological use of computer simulation, which is not to directly infer the past behaviour in question, but to test the efficacy of other methods of analysis. The role of such 'tactical' simulations is to provide data whose origin is understood in order to test the inferential power of the analytical technique in question (Orton 1982). Examples to date include tests of measures of the quantity of pottery (ibid.), multivariate statistics (Aldenderfer 1981), cladistic methods (Eerkens et al. 2005) and the relationship between temporal frequency distributions and prehistoric demography (Surovell and Brantingham 2007).

1.3.2 The Conceptual Accessibility of Agent-Based Modelling

It can be argued that specifically agent-based modelling is more accessible than other forms of computer simulation for both technical and conceptual reasons. The technical (i.e. computer science/software) dimension of agent-based modelling is discussed in Chaps. 3, 4 and 6, so here I focus on four ways in which agent-based modelling can be considered conceptually more accessible than other forms of computer simulation (such as dynamical systems modelling). Two of these concern the description of rules and outputs, and the other two concern what can be modelled.

First, it seems appropriate to start this discussion with the output of agent-based modelling, since that probably influences how agent-based modelling is perceived in the wider community. While it is true that the initial output from an agent-based model can be quite undigestable, comprising nothing more than many thousands of lines of a numerical log, it can be—and often is—visually attractive and, in particular, comprehensible in commonsense terms. At one extreme, it is now possible to integrate agent-based modelling with virtual reality so as to produce a photorealistic rendering of agents moving through a reconstructed palaeolandscape. This has been demonstrated for an agent-based simulation of Mesolithic hunter-gatherers foraging in a landscape that is now submerged under the North Sea (Ch'ng 2007; Ch'ng et al. 2011) and, while it is not clear what data is recorded for more conventional analysis, there is no doubt that this kind of realism captures the imagination of the public and less quantitatively inclined archaeologists alike. More common, however is two-dimensional mapping, for example, of the location of households presented as a series of snapshots at fixed intervals (e.g. Kohler et al. 2005, p. 79), the path followed by an individual agent (e.g. Costopoulos 2001, Fig. 9), or the distribution of artefacts discarded by agents (e.g. Lake 2000, Fig. 7). In my experience, non-specialists find the spatial patterns often output by agent-based simulations more readily comprehensible than the purely temporal patterns output by other methods such as dynamical systems models (perhaps with the exception of population curves, such as those found in Kohler et al. (2012a)). Moreover, the conclusions of dynamical systems models are often supported by phase portraits and phase diagrams (e.g. McGlade 1997), both of which assume a facility with mathematical abstraction that is not widespread among conventionally trained archaeologists. Admittedly, the output from some agent-based models is not so dissimilar (see Lake and Crema (2012) for an example) and it can certainly be argued (Premo 2007; also discussion in Sect. 1.5.1) that more agent-based models should be published with phase diagrams denoting the results of systematic experimentation, but the overarching point stands: even if there is no necessary connection between the visual output of a model and its scientific utility, agent-based simulations can and frequently do capture the imagination of non-modellers in a way that other simulation techniques do not.

Second, moving backwards from outputs to inputs, agent-based modelling is also accessible in the sense that it does not enforce one particular way of specifying the rules which govern the behaviour of entities in the model: this can be done mathematically, but it can also be done purely algorithmically (e.g. Rubio-Campillo et al. 2011; Costopoulos 2001) or using artificial intelligence formalisms such as production rules (e.g. Doran et al. 1994; Reynolds 1987). The fact that agent-based modelling does not require the mathematical formality of dynamical systems modelling may well partly account for its greater uptake, since algorithmic rules are typically closer in form to the verbal specifications of informal models.

Third, agent-based modelling is accessible in the sense that it offers great flexibility with respect to *what* can be modelled. It has already been noted that while most agent-based models share certain paradigmatic features, not all these are necessary and others can be added. Thus, for example, agent-based models can employ models of space ranging from purely topological networks to geographically referenced representations of the earth's surface. Similarly, agents may be purely reactive (they behave) or they may be cognitive (they reason using their own models of what is in their environment), and they may or may not communicate with other agents (Ferber 1999). Section 1.2 of this chapter listed some examples which demonstrate how this flexibility allows archaeologists with quite different interests to make use of agent-based modelling.

Finally, although agent-based modelling is—as a technique—scale agnostic, in most archaeological examples the agents are ethnographic-scale, in other words, they represent individual human beings or small groups, such as households. In practice, then, agent-based simulation is accessible in the sense that it tends to realign computer simulation with both the commonsense lay notion that archaeology is about what people did in the past, as well as the more explicitly theorised objections to systemic thinking raised by proponents of post-processual archaeology (e.g. Shanks and Tilley 1987b). The interweaving of archaeological computer simulation and archaeological theory is teased out in reviews by Aldenderfer (1998), McGlade (2005) and Lake (2014), but particular points of contact between post-1980 archaeological theory and agent-based modelling include agreement that "historical processes involve the actions of self-aware individuals" (Kohler 2000, p. 3) and the possibility of explicitly modelling cognition (Biskowski 1992; Doran 2000; Mithen 1989). That said, the detail of how agent-based models represent individuals and their relations with one another raises questions about the kind of rationality employed by agents (Lake 2004) and the locus of causality in human societies (Beekman 2005), and it is clear that some post-processual archaeologists (e.g. Thomas 1991, 2004) have not been won over on either count. Their concern, "Is is the world really agent-based?" (O'Sullivan and Haklay 2000), is discussed in more detail in Sect. 1.4.

1.3.3   Agent-Based Modelling as a Scientific Paradigm

As just suggested, there are clear points of contact between agent-based modeling and important strands in post-1980 archaeological theory. This is very conspicuous in some studies, notably for example Mithen's *Thoughtful Foragers*, which provides a book-length manifesto for a cognitive-processual archaeology (Renfrew 1994), in this case grounded in behavioural ecology and realised with the aid of simulation. Mithen's focus on individual decision-making went some way to addressing post processual theorists' (see Dornan (2002, pp. 308-314) for a review of the various positions taken) concern that the systemic perspective offered by processual archaeology overlooked the importance of individual agency and cognition (but see Sect. 1.4.1), but at the same time, however, he retained a firm commitment to scientific inference by hypothesis-testing. Nevertheless, although the commitment to scientific inference and a broadly evolutionary approach is characteristic of the vast majority of agent-based models (Lake 2014), relatively few share both the ethnographic-scale and realism (in the sense of attempted closeness to actual human decision-making and choice of parameter values) of Mithen's model. Instead, most archaeological agent-based models are either more abstract, or are based on higher-level agents (such as households); they are also frequently concerned with longer-term change and the emergence of new phenomena. All this suggests that the uptake of agent-based modelling is better explained by archaeologists' interest in the wider scientific paradigm which spawned it rather than its fit with post 1980 *archaeological* theory per se.

As with the geographical information systems (GIS) 'tool or science' debate (Wright et al. 1997), it is important to recognise that, as a technique, agent-based modelling can be used for different purposes and in a variety of theoretical frameworks. Nevertheless, agent-based modelling is closely associated with complexity science (Waldrop 1992),[1] whereas, for example, dynamical systems modelling is usually associated with chaos (Schuster 1988) and catastrophe theory (Zeeman 1977). Complexity science gained real momentum in the late 1980s/early 1990s and, although it builds on insights won by the formalisation of chaos and catastrophe in the 1960s and 1970s, it also marks a change of emphasis (Beekman and Baden

---

[1] Although agent-based modelling also has semi-independent roots in ecological modelling, where, as *individual-based* modelling the initial focus was on the importance of organism heterogeneity and spatial localism (DeAngelis and Gross 1992, p.xv).

2005). The mathematics of chaos provided tools to deal with the unpredictability and non-linearity of many real world processes, including some of obvious archaeological interest such as population growth (May 1976). Catastrophe theory ultimately proved more difficult to apply in the real world, but at minimum it reinforced the message that profound transformation does not necessarily require many factors working in tandem and nor does it necessarily require a strong external push (Renfrew and Poston 1979). Complexity science blends these insights with others from a different intellectual lineage that includes von Neumann's work on self-reproducing automata, cybernetics and connectionist cognitive science (Epstein and Axtell 1996, p. 2) to focus on the emergence of macro-level properties from the mutual interaction of many micro-level parts, as well as the related question of how such systems learn (adapt). Unlike dynamical systems models, which normally work with variables representing aggregate phenomena (e.g. the number of individuals who have access to prestige goods—see McGlade (1997)), agent-based models explicitly model the micro-level parts (as agents) and so are particularly well-suited to the complexity science agenda. Cellular automata share this particular property (the cells representing micro-level parts) and have also played a significant part in the development of complexity science (Toffoli and Margolus 1987; Wolfram 1984), but have found less use in archaeology as they are more restrictive in terms of the kinds of real-world phenomena they can represent (but see Mithen and Reed (2002) and Nikitas and Nikita (2005) for two archaeological applications of cellular automata).

The clearest manifestation of the impact of complexity science and agent-based modelling in the social sciences is the attempt to do what Epstein and Axtell (1996, p. 177) have labelled *generative social science*. Their book *Growing Artifical Societies* demonstrates how agent-based modelling can be used to grow social phenomena *in silico* "from the bottom up". In the most basic version of their now famous model Sugarscape, the actions of agents pursuing individual goals (to harvest and consume 'sugar' deposited on a landscape) produce population-level phenomena such as a characteristic power-law wealth distribution and waves of advance across space. Following further experiments with versions of Sugarscape that include more 'human' elements, including sexual reproduction, cultural transmission, combat, exchange and disease, Esptein and Axtell conclude that "A wide range of important social, or collective, phenomena can be made to *emerge* from the spatio-temporal interaction of autonomous agents operating on landscapes under simple local rules" (ibid., p. 153, my emphasis). Not only does this observation align social science with the interests of complexity science, but it also led Esptein and Axtell to propose a new model of explanation for the social sciences, one which they claim is neither deductive nor inductive, but 'generative': they interpret the question "can you explain it?" as asking "can you grow it?" (ibid., p. 177) and thus they propose that explanation consists in "demonstrating that certain sets of microspecifications are *sufficient to generate* the macrophenomenon of interest" (ibid., p. 20, original emphasis). On top of this epistemic move, Epstein and Axtell also argue that agent-based modelling provides the appropriate tool to overcome several specific deficiencies of contemporary social science: the use of local rather than global rules provides a "natural methodology for relaxing…assumptions" about the perfect rationality of actors (ibid., p. 1); the possibility of agents having different characteristics overcomes the explicit or implicit suppression of agent heterogeneity; and the focus on dynamics overcomes preoccupation with static equilibria at the expense of exploring transitional states which might actually be more important or interesting (ibid., p. 16). Overall, then, Epstein and Axtell hope that the agent-based approach will "yield *a new, more unified and evolutionary social science*, one in which migrations, demographic patterns, tribes, and tribal conflict, epidemics, markets, firms, institutions, and governments all emerge from the bottom up" (ibid., p162, original emphasis).

The Sugarscape agent-based model has been hugely influential: it is widely used for teaching computational social science and, within archaeology, is explicitly acknowledged as the inspiration for early examples of (modern) agent-based modelling such as the Long House Valley model (Axtell et al. 2002). Perhaps even more importantly, however, the very idea of generative social science has been gaining traction in archaeology. This is most obvious in writing that explicitly invokes *Growing Artificial Societies*, for example Kohler's (2000) manifesto for the potential of agent-based modelling to help with the task of "putting social sciences together again" and Premo's (2008, p. 36) call to use agent-based models as "behavioural laboratories", but it also underwrites other statements of the principal that agent-based models should be generative (e.g. Costopoulos 2009, p. 273). In fact, it can be argued that the majority of recent archaeological agent-based simulations adopt at least some aspects of the generative social science programme even if they do not explicitly invoke it. For example, many of the agent-based models designed to tackle problems in human evolution and/or evolutionary archaeology are concerned with emergence, the value of null models, or the less than perfect rationality of agents.

Premo's (2010) investigation of how a 'scatter and patches' archaeological landscape could have been produced by relatively simple Plio-Pleistocene hominin foraging and food-sharing behaviours is partly a study of an emergent phenomena. One could also argue that the models designed to demonstrate the effect of population size and structure on cultural trait diversity and cumulative innovation (e.g. Lake and Crema 2012; Powell et al. 2009; Premo 2012) also seek to explain how a population-level phenomenon, in this case cultural complexity, emerges from the interaction of agents.

Closely related to the notion of emergence is the realisation that complex patterns can be produced by the iteration of simple rules, which has in turn lead to recognition of the value of null models as a starting point of enquiry. Premo's aforementioned model of Plio-Pleistocene food-sharing is explicitly offered as null model (Premo 2007, p. 34). More recently, Bentley and Ormerod have argued for the utility of models which assume "zero-intelligence" (2012, pp. 205–6) on the part of agents. Bentley and Ormerod do not think that real-world agents really have zero-intelligence, but given that complexity science suggests that "many of the emergent, often complex, patterns in society need not require complex behavior on the part of individuals" (ibid., p. 205) they want to know how far we can get with simple social physics null models and what must be added to them to explain social phenomena. Bentley and colleagues have already demonstrated that a simple model of random copying is sufficient to explain the frequency distributions of cultural variants in first names, archaeological pottery, applications for technology patents, chosen dog breeds and popular music (Bentley et al. 2004, 2007).

As well as demonstrating the potential power of null models, the success of random copying models also reinforces our understanding that humans are not, in the main, perfectly rational decision-makers. In the context of null modelling this claim is usually predicated on the observation that humans do not have sufficient computational capacity to make optimal decisions and/or they do not have access to all the relevant information (e.g. Bentley and Ormerod 2012, pp. 205–6). Exactly what is meant by rationality is further discussed in Sect. 1.4.1, but the notion that humans are imperfect decision-makers because they only have access to spatially and/or temporally 'local' information has long been incorporated into archaeological agent-based models of hunter-gather foraging (see for example Reynolds 1987 and Mithen 1990); it is also now the explicit focus of enquiry in studies that seek to investigate the importance of noise (imperfect environmental tracking) in long-term adaptation (Xue et al. 2011).

Models such as those just cited demonstrate that the post-2000 uptake of agent-based modelling in archaeology has been accompanied by an interest in at least some strands of complexity science in general and generative social science in particular, even if it has not always been explicitly framed in those terms. While there are models which do not fit this characterisation so neatly, particularly tactical models (e.g. Rubio-Campillo et al. 2011), in the main there is a sense in which agent-based modelling has been adopted by archaeologists as both a technique *and* a scientific paradigm. It can be argued, however, that the particular strands of complexity science commonly found in archaeological agent-based models, such as relatively 'simple' emergent phenomena, the use of null models and a focus on limited rationality, reflect a particular perspective on complexity science, the Complex Adaptive Systems approach, that was strongly promoted by the Sante Fe Institute in the late 1980s and early 1990s. As Beekman and Baden (2005, p. 7) rightly caution, the application of ideas from complexity science to social phenomena has a different emphasis in Europe, where there is greater interest in explicitly cognitive issues, including emotion and irrationality (Doran 2000) and, perhaps most importantly, a greater willingness to treat social systems as qualitatively different from other complex systems (Gilbert 1995). Although they note that "archaeologists working within this intellectual milieu have often chosen to use nonlinear concepts as metaphors to frame a verbal analysis rather than develop computer simulations" (ibid.) it is worth considering whether this is because textbook agent-based modelling, which is largely grounded in the Complex Adaptive Systems approach, carries with it particular—and perhaps debatable—assumptions about the way the world is?

1.4   Ontological Considerations: Emergence and Rationality

As briefly alluded to above, the rapid deployment of GIS in a number of disciplines eventually led to debate about whether the technique is simply a tool that can be used for many purposes in a variety of theoretical frameworks, or whether it carries certain theoretical 'baggage' such that its use requires explicit or implicit adherence to particular theoretical principles or assumptions (in geography see Pickles (1999), Wright et al. (1997); in archaeology see Wheatley (1993), Gaffney and van Leusen (1995)). Similar questions have been asked of agent-based modelling, although the anxiety seems less widespread and largely confined to two key issues: is the rationality of agents a good model of human rationality and in what sense do social phenomena emerge.

1.4.1   The Rationality of Agents

As has been documented elsewhere (Lake 2004) and remains true today, the agents in archaeological agent-based simulations have generally been ascribed a rationality that is grounded in either human biology (typically optimal foraging theory or evolutionary psychology), or modern economics. Some archaeologists have argued that the biological and/or economic grounding of agent rationality is undesirable as it projects modern rationality back into the past and precludes the possibility of discovering that the past really was different. Thus, directly addressing agent-based modelling, Thomas (1991, p. 1) claimed that "The rationality which Mithen seeks to identify on the part of his Mesolithic foragers is a very specific one: it is the instrumental reason of late capitalism". In similar vein, but targeted at evolutionary archaeology more generally, Shanks and Tilley (1987a, p. 56) expressed disquiet about recourse to either evolutionary biology or economics, since in their view the former leaves a "plastic, malleable cultural dope incapable of altering the conditions of his or her existence" (Shanks and Tilley 1987a, p. 56) and the latter "naturalizes what are historically and culturally specific values as universal features of humanity" (Shanks and Tilley 1987b, p. 188). The cultural specificity of rationality has also been emphasised by Cowgill, who claimed that "The allegedly universal rationality assumed by 'economic man' models is shown by anthropological knowledge to be the very opposite" (2000, p. 55), and Clark who, comparing an approach to agency rooted in evolutionary theory with one rooted in practice theory, complained that "the individuals in optimal foraging models know more than real agents could know. Rational decisions require perfect knowledge of particulars and decision-making rules, which are cultural" (2000, p. 108).

The central question here is not whether attributing to agents rationalities grounded in evolutionary biology or economics is appropriate (see Kohler (2000) and Mithen (1990) for arguments in favour of evolutionary biology as a source of agent rationality), but rather, whether agent-based simulation is limited to these kinds of rationality? In order to address this question, however, it is necessary to be clear what one means by rationality. The textbook view from artificial intelligence is that "A rational agent is one that does the right thing" (Russell and Norvig 2003, p. 34), in other words, one "whose actions make sense from the point of view of the information possessed by the agent and its goals (or the task for which it was defined)" (Russell 1999, p. 13). The important and perhaps surprising consequence of this (informal) definition is that "What counts in the first instance is what the agent does, not necessarily what it thinks, or even whether it thinks at all" (*ibid*, p. 13). Thus the emphasis is first and foremost on what Simon (1956) termed *substantive rationality*—what decision to make, rather than *procedural rationality*—how to make the decision. For the purposes of computation, substantive rationality is formalised through the *agent function*, which can be conceived as a table that records what action an agent performs as a result of a given percept sequence (a history of everything the agent has ever perceived). Procedural rationality is formalised through the *agent program*, which is the internal mechanism used by the agent to implement the agent function and which, in the case of a cognitive agent, will be some kind of reasoning process.

Archaeologists who criticise the rationality accorded to agents in agent-based models and/or evolutionary and computational approaches more generally are not always explicit about what they mean by rationality, or at least not in terms that allow direct comparison with the way rationality is understood in artificial intelligence. The necessary translation requires a detailed textual analysis, which falls outside the scope of this paper but I have attempted it elsewhere (Lake 2004, pp. 195–197). I concluded that some critics (e.g. Clark 2000) are primarily concerned that the rationalities of past agents might need to be described by different agent functions, or as Clark put it, past agents "just have different motives" (ibid., p. 101). Where that is the concern, there are no grounds for rejecting agent-based modelling since the technique does not *require* agent functions derived from either modern economic theory or evolutionary biology. Of course, it must be possible to state the agent function in a computer programming language, so in practice it may be easier to do this for economic and biologically grounded rationalities for the simple reason that we have explicitly theorised them, but in principle, there is no reason why agent-based simulations should necessarily project modern *substantive* rationality back into the past. On the other hand, other critics of computational and evolutionary archaeology (Cowgill 2000; Thomas 1991) extend their concern to the agent program, that is, how agents make decisions: for example, Thomas states that "Desires, emotions, *forms of reason* and techniques of self-interpretation are all contingent and historically situated" (Thomas 1991, p. 17, my emphasis). Whereas one can in principle implement a wide range of agent functions in an agent-based model, it is less certain that agent programs are similarly unconstrained. Given that philosophers and artificial intelligence researchers are unable to agree about the limitations of machine intelligence (see Russell and Norvig (2003, chapter 26) for a guide to the main arguments), it is simply not possible to guarantee that all forms of reasoning that humans have ever employed can be implemented on the computing devices currently available to us, never mind in an agent-based model implemented using the standard software tools. However, it is also notable that archaeologists who warn against projecting modern forms of reasoning back into the past do not themselves provide detailed descriptions of alternative forms which they believe were employed in specific contexts, quite possibly because they do not have a language adequate to that task. Consequently, while it may be that agent-based simulation imposes some (unknown) limits on models of past *procedural* rationality, an attempt to devise non-modern agent programs would at least provoke much-needed discussion about how to describe alternative forms of reasoning.

So far I have considered the kind of rationality exhibited by agents, but Thomas' critique of Mithen's agent-based model of Mesolithic hunting as having created a "cybernetic wasteland" (1988, p. 64) extends to the argument that the agents in Mithen's model are unhuman because they are only rational and lack emotion. Mithen responded by acknowledging that emotional life is indeed "quintessentially human" (1991, p. 10), but largely because it is wrong to counterpose emotion and reasoning, since emotions actually make rational thought possible by overcoming the problem of what to do in the face of conflicting goals and/or the impossibility of predicting the consequences of all available actions. Effectively, by drawing in this way on the cognitive and functional theories of emotion proposed by Oatley and Johnson-Laird (1987) and Frijda (Frijda 1987; Frijda and Swagerman 1987), Mithen simply proposes that emotions are *implicitly* included in models of adaptation via rational decision-making (Mithen 1991, p. 9). Thus, while he argues that changes of state between, for example, stalking prey and killing it "are only made possible by emotions acting as a system of internal communication: monitoring the success and failure of sub-goals and the need to adjust behavioural plans" (ibid., p. 13), it is not the case that any of the computer code in his agent-based model could have been omitted had he denied any role for emotion.

In partial contrast to Mithen's approach to the problem of incorporating emotion into agent-based modelling, there is a strand of artificial intelligence research which seeks to very explicitly model the role of emotion in cognition (Doran 2000). This research is still grounded in the 'functional view' (Frijda 1995) that emotions serve to aid decision-making (Cañamero and de Velde 2000, p. 144), but in its most developed form it involves programming agents with a 'synthetic physiology', which includes variables necessary for survival (e.g. heart-rate, energy, blood sugar level) and hormones released under different emotional states which modify the amount of the controlled variables, thereby amplifying motives and thus ultimately influencing behaviour (Cañamero 1997). Thus, whereas the outcome of the emotional influence on rationality is completely predetermined in Mithen's model, this is not true in Cañamero's model, at least to the extent that the complexity of the interplay between emotions and reason may be sufficient to render it unpredictable in practice. If models like this really do offer the prospect of observing emotionally influenced behaviour that had not been *explicitly* imposed by the modeller then they open up the possibility of incorporating the role of emotion into the programme of generative social science. Indeed, Cañamero and de Velde (2000, p. 148) describe at length a conceptual design for an agent-based model which would allow each agent some control over the expression of emotions according to its state, interests and the image it has of the other, so allowing emotional states to contribute to the construction of intersubjectivity (ibid., p. 147). Although the bulk of research on agent emotion is situated within robotics (e.g. papers in Cañamero and Aylett (2008)), the fact that at least some artificial intelligence researchers are prepared to entertain a social constructivist view of emotions (Averill 1990) makes it difficult to imagine what *in-principle* objection remains to the use of agent-based simulation to model human decision-making.

1.4.2 The Emergence of Social Phenomena

As already noted, emergence is a central concept in complexity science and it is certainly the case that archaeologists have explicitly suggested that one of the benefits of using agent-based modelling is that it offers a means to study emergent phenomena. For example, Kohler and van der Leeuw suggest that agent-based models "enable us to examine the possibility of the emergence of new structures (for example, institutions, alliances, and communities) out of the basal units and their interactions" (2007a, p. 6) and go on to emphasise the possibility of modelling the recursive relationship between the emergent structures and the underlying micro-level entities such that "the interaction between individuals create the society (and its culture), which in turn, affect the behaviour of the individuals or groups that constitute it" (ibid., p. 7). Kohler also made a more specifically archaeological case for agent-based modeling: that the things we can measure in the archaeological record are on the one hand the outcome of agent behaviour "averaged over a great deal of space and time" but, given that, are also part of the "[context] within which agents make decisions and perform actions" (Kohler 2000, p. 10). This again points to the idea that agent-based modelling can be used to disentangle an element of recursion, in this case between agents and their environment. It was suggested above that a significant number of archaeological agent-based models have been used to investigate emergence in some sense, but the key question here is in what sense exactly? This is a difficult question to answer because there is no widely accepted formal theory of emergence (Epstein and Axtell 1996, p. 35) and indeed some would argue that the notion of emergence at best "remains vague and ill-defined" (Conte and Gilbert 1995, p. 9) and at worst "seems opaque, and perhaps even incoherent" (Bedau and Humphreys 2008, p. 1).

   In philosophical debate the concept of emergence becomes entwined with fundamental problems such as the mind–body problem (Searle 1992), but in complexity science, and especially agent-based modelling, the term 'emergence' is most often used "to denote stable macroscopic patterns arising from the local interaction of agents" (Epstein and Axtell 1996, p. 35). In the scheme proposed by the philosopher Bedau, cases of this kind of emergence are examples of either *nominal emergence* or *weak emergence*. Nominal emergence is the "simplest and barest" notion of emergence (Bedau 2008, p. 158), one in which the emergent phenomena have the kinds of properties that cannot be micro properties. Thus, for example, Epstein and

Axtell argue that the skewed wealth distribution produced by the Sugarscape model is an emergent phenomenon because "we do not know what it would mean for an agent to have a wealth distribution" (1996, p. 35). In similar vein, it is possible to measure the pressure of a gas, but not the pressure of an individual molecule of that gas. The difference between nominal and weak emergence is that in the former the emergent properties can be predicted and explained from the properties of the individual micro-level entities, whereas in the case of weak emergence "the micro-level interactions are interwoven in such a complicated network that the global behaviour has no simple explanation" (Bedau 2008, p. 160). In general, most of the agent based models developed under the banner of generative social science exhibit weak emergence. This is true of the archaeological models of Lake and Crema (2012), Powell et al. (2009) and Premo (2010, 2012) that were discussed earlier, and Beekman (2005, p. 64) provides another list which supports this observation. Most of these models effectively demonstrate—whether explicitly or implicitly—that "many of the emergent, often complex, patterns in society need not require complex behavior on the part of individuals" (Bentley and Ormerod 2012, p. 205), but on the other hand the interaction between individuals is often sufficiently complex (typically context dependent in some way) that it is not possible to predict the emergent phenomena except by running the simulation. This, of course, is what led Epstein and Axtell (1996, p. 177) to argue that adequate explanation of many social phenomena consists in demonstrating the "generative sufficiency" of a model (but see Sect. 1.5.1) for doubt about the adequacy of this proposition).

Despite the weight of actual modelling practice, there is considerable debate even within the modelling community about whether important social phenomena such as social institutions are the result of something more than weak emergence (Beekman 2005; Conte and Gilbert 1995; Gilbert 1995; Lake 2010; O'Sullivan and Haklay 2000). The two main issues are, first, whether at least some emergent social phenomena exert 'reverse' causal influence on agents (what Bedau labels *strong emergence*) and, second—but less discussed—whether it matters that human agents "are capable of reasoning, and do so routinely, about the emergent properties of their own societies" (Gilbert 1995, p. 144).

The reality of reverse causal influence is a long-standing point of contention in sociology (see Gilbert (1995) for more detail). On the one hand, methdological holists follow in the footsteps of Durkheim and Parsons in according causal influence to supra-individual social entities. This could hardly be clearer than in Durkheim's definition of social facts as "ways of acting, thinking and feeling that are external to the individual and are endowed with a coercive power by virtue of which they exercise control over him" (Durkheim 2004, p. 59). In similar vein, but much more recently, Kenneth Arrow argued that "social categories are used in economic analysis all the time and that they appear to be absolute necessities of the analysis, not just figures of speech that can be eliminated if need be" (1994, p. 1). In contrast, methodological individualists (e.g. Watkins 1952; O'Meara 1997, Mithen 1990 in archaeology) take the view that "large-scale social phenomena must be accounted for by the situations, dispositions and beliefs of individuals" (Watkins 1952, p. 58) and do not accept the ontological reality of social institutions, since even social roles "can be fully understood in terms of individuals as long as we take a wide enough perspective so as to include all those individuals, in other times and places, who create a role" (King 1999, p. 216). The last three decades of the Twentieth Century saw numerous attempts to negotiate or even dissolve the micro-macro dichotomy in sociology (Beekman 2005, p. 53). The most frequently cited in archaeology are Gidden's (1984) Structuration Theory and Bourdieu's (1977) Practice Theory, but in an article specifically examining the relationship between social theory and agent-based simulations, Beekman notes that the tendency among archaeologists to chose one or the other—along with the particular readings that have found their way into archaeology—has in practice ended up replicating the dichotomy between discursive strategizing action and non-discursive practice that characterises the debate between methodological individualists and methodological holists (Beekman 2005, p. 55). Noting that even individual human beings are emergent entities, Beekman's own preference is an approach grounded in—but not slavishly adhering too—Archer's (2000) critique of Structuration Theory (Beekman 2005, pp. 62–3), which recognises multiple scales of collective "at which different rules of action predominate". He considers that collective agents are "real social entities" (ibid., p. 68) and although he does not explicitly say so, strongly implies that these entities exert reverse causal force on lower-level agents.

This is not the place to attempt to adjudicate between methodological holism, methodological individualism and attempts to overcome that dichotomy, but it is worth reiterating the point that actual practice in agent-based modelling tends towards methodological individualism (Beekman 2005; O'Sullivan and Haklay 2000). Even the few archaeological agent-based models that have explicitly modelled both individuals and groups (see Doran et al. 1994; Kohler et al. 2012b; Lake 2000) do not really exhibit the *emergence* of reverse causal force since in all cases the establishment of groups is scaffolded at the outset of the simulation. Consequently, it can be argued that, to date, the practice of archaeological agent-based simulation does not match Kohler and van der Leeuw's ambition to investigate the recursive relationship between individuals and society. Given the lack of unanimity among sociologists about the status of social institutions it is far from clear whether this really matters, especially given philosophical doubts about the ontological status of reverse causal force (e.g. see Bedau (2008, p. 159), who argues that "strong emergence starts where scientific explanation ends"). That said, Gilbert makes the important point that if the definition of weak emergence hinges on the impossibility of analytically predicting the macro-level phenomena then "any particular emergent property stands the risk of being demoted from the status of emergence at some time in the future" (1995, p. 150), which leads him to suggest that the relationship between micro and macro properties of complex systems may be more interesting than emergence per se. Following in this spirit there may at the very least be scope to use agent-based modelling to develop a better understanding of whether or how reverse causal force amounts to more than "feedback effects in the agent population, altering the behaviour of individuals" (Epstein and Axtell 1996, pp. 16–7).

The second—albeit less discussed—issue in the debate about whether important social phenomena such as social institutions are the result of something more than weak emergence is the question of whether it matters that human agents reason about the emergent properties of their own societies. Some taxonomies of emergence do indeed distinguish between emergence that involves a degree of reflexivity and emergence that does not. For example, cognitive psychologist and artificial intelligence researcher Cristiano Castelfranchi has proposed four senses of emergence, of which the third is *representational*, meaning that the phenomenon in question is "learned or recognized at the cognitive level" (Conte and Gilbert 1995, p. 9). In not dissimilar vein, Sunny Auyang posits a three level hierarchy of complex systems in which the third is cybernetic systems that involve intentionality (Auyang 1998). Sociologist Nigel Gilbert clearly believes that failure to take account of the capacity of humans to "perceive, monitor and reason with the macro-properties of the society in which they live" (1995, p. 155) undermines the utility of simulation for addressing the kinds of questions of interest to his discipline.

One potential difficulty with the argument that the reflexivity of humans gives rise to a special kind of emergence is that, although the human capacity to be reflexive is not contested, there is debate—including within archaeology (see Hodder 2001, p. 10)—about the extent to which social practices are undertaken by agents who have practical knowledge, that is, who know "how to go on" in the world "as it is" without consciously reflecting upon it (Barrett 2001, p. 151). This raises the spectre of a quantitative gradation (how much reflection is taking place) underpinning a qualitative distinction (a special kind of emergence). I suggest that Crutchfield's (2008) concept of intrinsic emergence may provide a solution to this conundrum. Crutchfield notes that in well-known physical examples of self-organizing phenomena (e.g. Rayleigh-Bénard convection cells in heated liquids) "the patterns which appear are detected by the observers and analysts" but, crucially, there is no reason to believe that those experiments behaved differently before the self-organizing phenomena were detected by scientists (ibid, p. 271). Consequently, he argues that "it is the observer or analyst who lends the teleological 'self' to processes which otherwise simply 'organize' according to the underlying dynamical constraints" (ibid.). Crutchfield proposes that, in contrast, intrinsic emergence occurs when "the patterns formed confer additional functionality which supports global information processing" (ibid., p. 272), or in other words, the system itself has 'discovered' the pattern. Significantly for our purposes, however, Crutchfield makes it clear that 'discovery' does not require a cognitive representation, but can be "[implicit] in the dynamics and behaviour of a process" if the system makes use of the new pattern (ibid.). Thus, the concept of intrinsic emergence may provide a means of recognising the reflexivity of social systems (albeit not necessarily just human systems) as something more profound than weak emergence, but without running into the problem of measuring the extent to which social practice is practical or discursive.

1.5 Epistemic Considerations: How to Learn by Agent-Based Modelling

Earlier it was suggested that archaeologists use computer simulation for one or more of four purposes: enforcing conceptual clarity, helping understand how things change, helping infer past behaviour from a static archaeological record, and testing other quantitative methods. The second and third of these are of particular interest here, since models built for understanding how things change and those built for inferring past behaviour typically constitute the most fully worked through attempts to explain aspects of the human past. Within the archaeological simulation literature there is an established distinction between 'theory-building' and 'hypothesis-testing' models (e.g. Mithen 1994, pp. 176–177), according to which the aim of hypothesis-testing models is to determine what actually happened in the past by comparing the output of a simulated process against the archaeological evidence, while the use of simulation models to support theory-building does not necessarily involve detailed comparison of output against the archaeological record since the purpose is not to test what happened in the past, but rather to understand how certain processes work and what sort of changes could plausibly have occurred. It can increasingly be questioned whether this is a particularly useful distinction, since, on the one hand, comparing simulation output with the archaeological evidence can contribute to theory-building (e.g. Kohler and Varien 2012), while on the other hand, simulation can be used to directly test hypotheses which are more about possible processes (e.g. the effect of parameters on model dynamics) than what actually happened in the past (see Premo 2010, pp. 29–30). Rather than attempting to force models into a rigid taxonomy according to which each class represents a discrete inferential role, it seems more productive to focus attention on issues of modelling logic that commonly arise when attempting to learn by simulation.

1.5.1 Learning by Simulation Requires Experimentation not Just Emulation

Earlier it was noted that the visual accessibility of many agent-based models may well be a significant factor in the rapidly growing popularity of the technique. Para-doxically, however, this same accessibility to non-specialists may also contribute to a lack of understanding that effective use of simulation requires an experimental approach. In a series of articles, Premo (2005, 2007, 2008) has been particularly forceful in arguing that merely *emulating* the past does not *explain* it. The basic problem is relatively simple: by iteratively adjusting the process and/or parameters of a model it will probably be possible to obtain a reasonable fit between the model output and the archaeological evidence (an emulation), but this does not guarantee that only this process/parameters could have caused what happened in the past (the problem of equifinality) and nor does it provide a good understanding of the probability of this particular outcome versus others. Such *underdetermination* might not be a problem for those persuaded by constructive empiricism as an account of science, that is, the proposition that the aim of science is to produce theories that are empirically adequate rather than literally true (Fraassen 1980; see Kelley and Hanen (1988) for archaeological supporters), but it is a problem if one believes, as do most contemporary philosophers of science (Preston 2014), that scientific theories are literally either true or false, whether one can actually know that (as per scientific realism) or not (as per conjectural realism). It also casts some doubt on Epstein and Axtell's proposition that the programme of generative social science should equate explanation with answering the question "can you grow it?" (1996, p. 177), since successfully 'growing' some phenomenon does not automatically overcome the problem of underdetermination—that there might be some other way of growing the same phenomenon.

Given that the significance of underdetermination is a major point of contention in the philosophy of science, it is hardly surprising that there is no simple solution for overcoming the problem of equifinality in archaeological inference. That said, computer simulation at least has the advantage that, if used as a "behavioural laboratory" (Premo 2008), it allows us to explicitly explore what alternative models might equally or better fit the observed phenomenon, and/or do so with less sensitivity to aspects of the model for which there is limited independent evidence. This requires two important moves. The first is that modelling starts with an explicit theory, in order that the prior understanding is rendered brittle enough by the need to maintain internal consistency that it can be broken, that is, found to be inadequate. The second move is to adopt an experimental exploratory design (ibid., p. 49), one which does not simply attempt to replicate some observed phenomenon, but systematically explores the consequences of the model under a wide range of circumstances, only some of which may actually have obtained in the past. By re-running Gould's 'tape of history' (see Premo (2008, pp. 49–50) for detailed discussion) in this way, the modeller can generate an ensemble of "'what if' scenarios" (ibid., p. 50) or "alternative cultural histories" (Gumerman and Kohler 2001) which can then provoke rejection of the model or form the basis of an explanation of the observed phenomenon in the sense of delineating what conditions must have been met for that phenomenon to have occurred if the model is correct.

Although appropriate experimental design is vital to productive use of simulation, two intrinsic properties of simulation models are also important: whether a model is generative and whether it is simple or complex. Each is considered in turn.

1.5.2   The Most Informative Models Are Generative

As discussed in Sect. 1.3.3, the program of generative social science is built around a particular view of what constitutes an explanation. For Epstein and Axtell the aim of generative social science is "to provide initial microspecifications (initial agents, environments, and rules) that are *sufficient to generate* the macrostructures of interest" and thus they "consider a given macrostructure to be 'explained' by a given microspecification when the latter's generative sufficiency has been established" (1996, p. 177). As noted above, it can be argued that this formulation takes insufficient account of underdetermination, but that does not detract from the point that explanation requires that the macrostructures of interest (e.g. social institutions or other population-level outcomes) must not have been programmed into the computer simulation at the outset, since were that the case then the model would simply describe a known outcome. As Costopoulos (2009, p. 273) puts it, the explanatory power of a generative model lies in the fact that it "must be observed in operation to find out whether it will produce a predicted outcome".

While the power of generative modelling is widely acknowledged among simulation-using archaeologists (e.g. Beekman 2005; Costopoulos 2009; Kohler 2000; Premo 2008), its use does raise a practical problem of system-bounding if one is to avoid infinite regress. Thus, for example, Beekman (2005, p. 66) has proposed that "the only rules that should be fixed within a simulation should be the most basic biological imperatives common to all humans, while any Giddensian structural rules and resources must emerge through agents' actions". While this may be appropriate if the purpose of the simulation is specifically to study how society 'works' (the context in which Beekman made this suggestion), it is less clear that one really needs to simulate the construction of the self and person *sensu* Simmel (see Scaff (2011, pp. 213–225) and Archer (2000)) in order to study, say, the effect of population size on the maintenance of cultural complexity, or how foraging in a patchy environment promotes food-sharing. Even sociologists who reject the ontological reality of social institutions accept that for practical purposes it may be necessary "to assume certain background conditions which are not reduced to their micro dimensions" (King 1999, p. 223). It is important to recognise, however, that under these circumstances one can not claim that the model in any way explains the assumed macro-level properties. Thus, for example, although Lake's (2000) model of Mesolithic foraging was one of the relatively few early archaeological agent-based models which explicitly incorporated 'social' behaviour above the level of the basic agent (in this case collective decision-making by groups of family units), the model sought only to explain spatial outcomes of that decision-making, not the emergence or manner of the decision-making itself (see also commentary in O'Sullivan and Haklay 2000, p. 1419). Ultimately, as Kohler and van der Leeuw remind us "A good model is not a universal scientific truth but fits some portion of the real world reasonably well, in certain respects and for some specific purpose" (2007b, p. 3). Consequently, a pragmatic stance is that what matters is not whether a model is maximally generative, but whether it is generative *with respect to its purpose*, that is to say, it must incorporate agent—agent and agent—environment interaction relevant to its scale and it must be necessary to run the model in order to find out whether the particular phenomena of interest emerges.

1.5.3   Explanatory Power Trades Complexity Against Fit

The issue of what is included in a model raises the question of whether models should be as simple as possible, or more 'lifelike'. As Levins (1966) long ago noted in the context of population modelling, it is in practice impossible to simultaneously maximise the generality, realism, and precision of models of complex systems. For some, this implies that choices must be made according to the intended scope and purpose of the model (Kohler and van der Leeuw 2007b, pp. 7–8), whereas others see a strong presumption in favour of simplicity (Premo 2008, p. 48).

There are three main arguments in favour of simplicity (which are not mutually exclusive). The first and most basic argument is that replicating a complex world by means of a complex model is unlikely to lead to enhanced understanding since the latter is achieved by reducing complexity to "intelligible dimensions" (Wobst 1974, p. 151). In other words, explanation requires reduction, although whether that is more due to the limitations of the human intellect than it is a reflection of the way the world is has long been debated (see discussion on this point in Laird 1919, pp. 342–4). Either way, Collard and Slingerland argue that in practice both scientists and humanists reduce, since "any truly interesting explanation of a given phenomenon is interesting precisely because it involves reduction of some sort–tracing causation from higher to lower levels or uncovering hidden causal relationships at the same level" (2012, location 311); indeed, they go so far as to suggest that "when someone fails to reduce we rightly dismiss their work as trivial, superficial, or uninformative" (ibid., location 314).

Not unrelated is the second argument in favour of simplicity: the application of the law of parsimony, which posits that one should adopt the simplest explanation for the observed facts. Indeed, one of the most important insights from complexity science is the discovery that complex macro-level patterns do not necessarily require complex behaviour on the part of individual agents. As noted earlier, this has lead to enthusiasm for null modelling (Premo 2007), in which one starts by investigating how much of the observed phenomenon can be explained by the simplest possible model. Although it could be argued that complexity science adopts an ontological stance in favour of the notion that complexity is generated by the interaction of agents which individually exhibit relatively simple behaviour, null modelling can also be viewed as an epistemic move favouring the gradual addition of complexity to models to establish if doing so allows them to explain more of the patterning in the observed data (Premo 2007, p. 34). Thus, for example, while Bentley and Ormerod argue "that the most appropriate 'null model' of individual behavior in larger societies is in fact... the 'zero-intelligence' model" (2012, pp. 205–6), they also write at length about what needs to be added to a null model based on statistical physics precisely because human interactions are different. This may seem obvious, but the point of modelling "from the null-up" (Premo 2007, p. 34) is to avoid making assumptions about complexity in favour of discovering how much complexity is necessary to explain the observed phenomenon. Kohler et al. (2012a, p. 40) make a very similar point in relation to the use of optimizing models when they state that they "do not... want to predetermine the answer to fundamental questions such as, 'do societies operate so as to optimize the actions of their members?'—since these are questions we would like to ask".

The third argument in favour of simplicity is the argument from generality (e.g. Costopoulos 2009, p. 275): that simpler models which have not been finely honed to fit a particular case, but can account for more—and more diverse—cases have greater explanatory power, not least because they allow one to predict what should happen in a wider range of circumstances and so obtain a greater sense of the likelihood of the observed phenomenon occurring rather than some other phenomenon. As Pinker forcefully argues, explanation requires more than "saying something just is": it consists in demonstrating "why it had to be that way *as opposed to some other way it could have been*" (Pinker 2002, p. 72, my emphasis).

Although there are strong arguments in favour of keeping models simple, a common view among philosophers of science is that "the best model for a given data set is one which balances order and randomness by minimizing the model's size while simultaneously minimizing the 'amount of apparent randomness' " on the grounds that such a model ensures that "causes [are not] multiplied beyond necessity while also obtaining a good prediction" (Crutchfield 2008, p. 274). In ecological modelling it is increasingly argued that the best way to find this "optimal zone of model complexity" is to build models that are structured to reproduce *multiple* real-world patterns, not least because it is thought that such models are usually less sensitive to parameter uncertainty (Grimm et al. 2005, p. 989; also Piou et al. 2009). This approach may also be profitable in archaeology (Altaweel et al. 2010), but it is not unreasonable so suppose that the optimal balance between model size and fit may vary within a discipline which encompasses such wide-ranging subject-matter studied at a variety of spatial and temporal scales. For instance, minimizing apparent randomness may be important in a model designed to investigate whether Mesolithic land-use patterns on a small Scottish island reflect the exploitation of specific resources (Lake 2000), but on the other hand a more appropriate ambition for an archaeology of the very long-term might be whether there is an "'envelope of predictability' for major socio-environmental changes, within which specific events and timings remain unpredictable?" (Cornell et al. 2010, p. 427). Ultimately then, it may be that models of varying simplicity and fit can be productive providing they meet two conditions: (i) they are generative with respect to the problem at hand; and (ii) they adopt an exploratory experimental design in order to elucidate other ways the explanandum could have been. In practice, such models are likely to be those at the simpler end of the spectrum, but this need not always be so.

1.6 Summary

Archaeologists have experimented with computer simulation for almost as long as they have used computers and even some of the earliest simulation models have features in common with contemporary agent-based models. Nevertheless, there has been an explosion of interest in agent-based simulation modelling since 2000, driven by its conceptual flexibility and accessibility, the appearance of relatively 'user-friendly' software and interest in the wider agenda of complexity science. Indeed it can be argued that the technique has now achieved a degree of maturity: its use in certain subdisciplines (e.g. evolutionary archaeology) is becoming literally unremarkable, such that papers increasingly focus on results and their implications for substantive problems rather than methodological issues. Even so, there is scope for greater consideration of what is required to maximise the potential of learning by simulation, particularly with regard to experimental design: ensuring that results are not 'built in' and achieving an appropriate balance between model complexity and the fit to data. Furthermore, there remain questions about what ontological baggage, if any, comes with the adoption of agent-based modelling. Many, if not most, archaeological agent-based models adopt a fairly strong methodological individualism and concomitantly weak notion of emergence. Is this why, or because, most archaeological agent-based models deal with small-scale societies? Is it just sensible scientific scepticism of mysterious downward causal forces, or is it a narrow-minded and premature closing down of the possibility of a scientific account of long-term social change? At present the answer is far from clear, but intelligent application of agent-based modelling to a more diverse range of problems will surely help to tease out what is required for satisfactory explanation of aspects of human history.

Acknowledgements I should like to thank Gabriel Wurzer and Kerstin Kowarik for inviting me to contribute to this volume and also for their hospitality in Vienna during the 2011 trans-disciplinary workshop *Agents in Archaeology*. I am also grateful to many others with whom I have debated the pros and cons of agent-based modelling and/or who have invited me to present or discuss agent-based models in various workshops and conferences over the years, notably: Ariane Burke, Mark Collard, Andre Costopoulos, Enrico Crema, Tim Kohler, Marco Madella, Steven Mithen, Luke Premo, Bernado Rondelli and James Steele.

References

Aldenderfer MS (1981) Creating Assemblages by Computer Simulation: The Development and Uses of ABSIM. In: Sabloff JA (ed) Simulations in Archaeology, University of New Mexico, Albuquerque, pp 11–49
Aldenderfer MS (1998) Quantitative methods in archaeology: A review of recent trends and developments. J Archaeol Res 6:91–120
Allen P, McGlade J (1987) Evolutionary drive: the effect of microscopic diversity. Found Phys 17:723–738
Allen PM, Strathern M, Baldwin JS (2006) Evolutionary drive: new understandings of change in socio-economic systems. Emergence 8(2):2–19
Altaweel M, Alessa L, Kliskey A, Bone C (2010) A framework to structure agent-based modeling data for social-ecological systems. Struct Dynamics 4(1):article 2, URL http://escholarship.org/uc/temporary?bpid=1061732
Archer M (2000) Being Human: The Problem of Agency. Cambridge University Press, Cambridge
Arrow K (1994) Methodological individualism and social knowledge. Am Econ Rev 84(2):1–9

Auyang SY (1998) Foundations of Complex-System Theories: In Economics, Evolutionary Biology, and Statistical Physics. Cambridge University Press, Cambridge

Averill JR (1990) A Constructivist View of Emotion. In: Plutchik R, Kellerman H (eds) Emotion: Theory, Research and Experience, vol 1. Academic Press, New York, pp 305–339

Axtell RL, Epstein JM, Dean JS, Gumerman GJ, Swedlund AC, Harburger J, Chakravarty S, Hammond R, Parker J, Parker M (2002) Population growth and collapse in a multiagent model of the Kayenta Anasazi in Long House Valley. Proc Natl Acad Sci USA 99(3):7275–7279. DOI 10.1073/pnas.092080799

Barrett JC (2001) Agency, the Duality of Structure and the Problem of the Archaeological Record. In: Hodder I (ed) Archaeological Theory Today. Polity Press, Cambridge, pp 141–164

Barton C, Ullah I, Mitasova H (2010a) Computational modeling and Neolithic socioecological dynamics: a case study from southwest Asia. Am Antiq 75(2):364–386. URL http://saa.metapress.com/index/1513054509884014.pdf

Barton CM (2013) Stories of the Past or Science of the Future? Archaeology and Computational Social Science. In: Bevan A, Lake M (eds) Computational Approaches to Archaeological Spaces Left Coast Press, Walnut Creek, CA, pp 151–178.

Barton CM, Ullah II, Bergin S (2010b) Land use, water and Mediterranean landscapes: modelling long-term dynamics of complex socio-ecological systems. Phil Trans Roy Soc A Math Phys Eng Sci 368(1931):5275–5297. URL http://rsta.royalsocietypublishing.org/content/368/1931/5275.short

Bedau MA (2008) Downward Causation and Autonomy in Weak Emergence. In: Bedau MA, Humphreys P (eds) Emergence: Contemporary Readings in Philosophy and Science. MIT Press, Cambridge, MA, pp 155–188

Bedau MA, Humphreys P (2008) Introduction. In: Bedau MA, Humphreys P (eds) Emergence: Contemporary Readings in Philosophy and Science. MIT Press, Cambridge, MA, pp 1–6

Beekman CS (2005) Agency, Collectivities and Emergence: Social Theory and Agent Based Simulations. In: Beekman CS, Baden WW (eds) Nonlinear Models for Archaeology and Anthropology. Ashgate, Aldershot, pp 51–78

Beekman CS, Baden WW (2005) Continuing the Revolution. In: Beekman CS, Baden WW (eds) Nonlinear Models for Archaeology and Anthropology. Ashgate, Aldershot, pp 1–12

Belovsky GE (1987) Hunter-gatherer foraging: A linear programming approach. J Anthropol Archaeol 6(1):29–76. URL http://www.sciencedirect.com/science/article/pii/027841658790016X

Bentley A, Ormerod P (2012) Agents, Intelligence, and Social Atoms. In: Collard M, Slingerland E (eds) Creating Consilience: Reconciling Science and the Humanities. Oxford University Press, Oxford, pp 205–222. URL http://books.google.com/books?hl=en&lr=&id=rERUMR0tek8C&oi=fnd&pg=PA205&dq=Agents,+intelligence,+and+social+atoms&ots=mexDG9rPbj&sig=w5Lq0JfO7-qaZDu4HTkmFnNjTSQ

Bentley R, Lipo C, Herzog H, Hahn M (2007) Regular rates of popular culture change reflect random copying. Evol Hum Behav 28(3):151–158. URL http://www.sciencedirect.com/science/article/pii/S109051380600095X

Bentley RA, Hahn MW, Shennan SJ (2004) Random drift and culture change. Proc Roy Soc Lond B 271:1443–1450

Bentley RA, Lake MW, Shennan SJ (2005) Specialisation and wealth inequality in a model of a clustered economic network. J Archaeol Sci 32:1346–1356

Binford LR (1981) Bones: Ancient Men and Modern Myths. Academic Press, New York

Biskowski M (1992) Cultural Change, the Prehistoric Mind, and Archaeological Simulations. In: Reilly P, Rahtz S (eds) Archaeology and the Information Age. Routledge, London, pp 212–229

Bourdieu P (1977) Outline of a Theory of Practice. Cambridge University Press, Cambridge

Cañamero D (1997) Modeling Motivations and Emotions as a Basis for Intelligent Behaviour. In: Johnson WL (ed) Proceedings of the First International Conference on Autonomous Agents. ACM Press, New York, pp 148–155

Cañamero D, de Velde WV (2000) Emotionally Grounded Social Interaction. In: Dautenhahn K (ed) Human Cognition and Social Agent Technology. John Benjamins Publishing, Amsterdam, pp 137–162

Cañamero L, Aylett R (eds) (2008) Animating Expressive Characters for Social Interaction. John Benjamins Publishing, Amsterdam

Ch'ng E (2007) Using games engines for archaeological visualisation: Recreating lost worlds. In: Proceedings of CGames 2007 (11th International Conference on Computer Games: AI, Animation, Mobile, Educational & Serious Games), La Rochelle, France (2007), vol 7, pp 26–30

Ch'ng E, Chapman H, Gaffney V, Murgatroyd P, Gaffney C, Neubauer W (2011) From sites to landscapes: How computing technology is shaping archaeological practice. Computer 44(7):40–46. URL http://ieeexplore.ieee.org/xpls/abs_all.jsp?arnumber=5871564

Clark JE (2000) Towards a Better Explanation of Hereditary Inequality: A Critical Assessment of Natural and Historic Human Agents. In: Dobres MA, Robb JE (eds) Agency in Archaeology. Routledge, London, pp 92–112

Clarke DL (ed) (1972) Models in Archaeology. Methuen, London

Conte R, Gilbert N (1995) Introduction: Computer Simulation for Social Theory. In: Gilbert N, Conte R (eds) Artificial Societies: The Computer Simulation of Social Life. UCL Press, London, pp 1–18

Cornell S, Costanza R, Sörlin S, van der Leeuw S (2010) Developing a systematic "science of the past" to create our future. Global Environ Change 20(3):426–427

Costopoulos A (2001) Evaluating the impact of increasing memory on agent behaviour: Adaptive patterns in an agent-based simulation of subsistence. J Artif Soc Soc Simulat 4. URL http://jasss.soc.surrey.ac.uk/4/4/7.html, http://www.soc.surrey.ac.uk/JASSS/4/4/7.html

Costopoulos A (2009) Simulating Society. In: Maschner H, Bentley RA, Chippindale C (eds) Handbook of Archaeological Theories. Altamira Press, Lanham, Maryland, pp 273–281

Cowgill GE (2000) "Rationality" and Contexts in Agency Theory. In: Dobres MA, Robb JE (eds) Agency in Archaeology. Routledge, London, pp 51–60

Crema ER (2013) A simulation model of fission-fusion dynamics and long-term settlement change. J Archaeol Method Theory. DOI: 10.1007/s10816-013-9185-4

Crutchfield JP (2008) Is Anything Ever New? Considering Emergence. In: Bedau MA, Humphreys P (eds) Emergence: Contemporary Readings in Philosophy and Science. MIT Press, Cambridge, MA, pp 269–286, originally published in Cowan, Pines and Meltzer eds, 1999, *Complexity: Metaphors, Models and Reality*, Westview Press

Dean JS, Gumerman GJ, Epstein JM, Axtell RL, Swedlund AC, Parker MT, McCarroll S (2000) Understanding Anasazi Culture Change Through Agent-Based Modeling. In: Kohler TA, Gumerman GJ (eds) Dynamics in Human and Primate Societies: Agent-Based Modelling of Social and Spatial Processes. Santa Fe Institute Studies in the Sciences of Complexity. Oxford Univesity Press, New York, pp 179–205

DeAngelis DL, Gross LJ (1992) Individual-Based Models and Approaches in Ecology: Populations, Communities and Ecosystems. Chapman & Hall, New York

Doran JE (1970) Systems theory, computer simulations, and archaeology. World Archaeol 1:289–298

Doran JE (2000) Trajectories to Complexity in Artificial Societies: Rationality, Belief and Emotions. In: Kohler TA, Gumerman GJ (eds) Dynamics in Human and Primate Societies: Agent-Based Modelling of Social and Spatial Processes. Oxford University Press, New York, pp 89–144

Doran JE, Hodson FR (1975) Mathematics and Computers in Archaeology. Edinburgh University Press, Edinburgh

Doran JE, Palmer M (1995) The EOS Project: Integrating Two Models of Palaeolithic Social Change. In: Gilbert N, Conte R (eds) Artificial Societies: The Computer Simulation of Social Life. UCL Press, London, pp 103–125

Doran JE, Palmer M, Gilbert N, Mellars P (1994) The EOS Project: Modelling Upper Palaeolithic Social Change. In: Gilbert N, Doran J (eds) Simulating Societies. UCL Press, London, pp 195–221

Dornan JL (2002) Agency and archaeology: past, present, and future directions. J Archaeol Method Theory 9(4):303–329. URL http://link.springer.com/article/10.1023/A:1021318432161

Durkheim E (2004) Readings from Emile Durkheim. Routledge, New York. Edited by Kenneth Thompson

Eerkens JW, Bettinger RL, McElreath R (2005) Cultural Transmission, Phylogenetics, and the Archaeological Record. In: Lipo CP, O'Brien MJ, Collard M, Shennan SJ (eds) Mapping Our Ancestors: Phylogenic Methods in Anthropology and Prehistory. Transaction Publishers, Somerset, NJ, pp 169–183

Epstein JM, Axtell R (1996) Growing Artificial Societies: Social Science from the Bottom Up. Brookings Press and MIT Press, Washington

Ferber J (1999) Multi-Agent Systems: An Introduction to Distributed Artificial Intelligence, english edn. Addison-Wesley, Harlow

Fraassen B (1980) The Scientific Image. Clarendon Press, Oxford

Frijda NH (1987) The Emotions. Cambridge University Press, Cambridge

Frijda NH (1995) Emotions in Robots. In: Roitblat HL, Meyer JA (eds) Comparative Approaches to Cognitive Science. MIT Press, Cambridge, MA, pp 501–516

Frijda NH, Swagerman J (1987) Can computers feel? Theory and design of an emotional system. Cognit Emot 1:235–257

Gaffney V, van Leusen PM (1995) Postscript—GIS, Environmental Determinism and Archaeology: A Parallel Text. In: Lock GR, Stančič Z (eds) Archaeology and Geographical Information Systems: A European Perspective. Taylor & Francis, London, pp 367–382

Giddens A (1984) The Constitution of Society: Outline of a Theory of Structuration. Polity Press, Cambridge

Gilbert N (1995) Emergence in Social Simulation. In: Gilbert N, Conte R (eds) Artificial Societies: The Computer Simulation of Social Life. U.C.L. Press, London, pp 144–156

Gilbert N (2008) Agent-Based Models. Quantitative Applications in the Social Sciences. Sage, Thousand Oaks, CA, URL http://books.google.co.uk/books?id=Z3cp0ZBK9UsC

Gould SJ (1989) Wonderful Life: The Burgess Shale and the Nature of History, paperback edn. Vintage, London

Grimm V, Railsback S (2005) Individual-Based Modeling and Ecology. Princeton University Press, Princeton

Grimm V, Revilla E, Berger U, Jeltsch F, Mooij WM, Railsback SF, Thulke HH, Weiner J, Wiegand T, DeAngelis DL (2005) Pattern-oriented modeling of agent-based complex systems: lessons from ecology. Science 310(5750):987–991. URL http://www.sciencemag.org/content/310/5750/987.short

Gumerman GJ, Kohler TA (2001) Creating Alternative Cultural Histories in the Prehistoric Southwest: Agent-Based Modelling in Archaeology. In: Examining the Course of Southwest Archaeology: The Durango Conference, September 1995, New Mexico Archaeological Council, Albuquerque, pp 113–124

Hodder I (2001) Introduction: A Review of Contemporary Theoretical Debates in Archaeology. In: Hodder I (ed) Archaeological Theory Today. Polity Press, Cambridge, pp 1–13

Innis GS (1972) Simulation of ill-defined systems, some problems and progress. Simulation 19:33–36

Johnson CD, Kohler TA (2012) Modeling Plant and Animal Productivity and Fuel Use. In: Kohler TA, Varien MD, Wright AM (eds) Emergence and Collapse of Early Villages: Models of Central Mesa Verde Archaeology. University of California Press, Berkeley, pp 113–128

Kachel AF, Premo LS, Hublin JJ (2011) Grandmothering and natural selection. Proc Roy Soc B Biol Sci 278(1704):384–391. URL http://rspb.royalsocietypublishing.org/content/278/1704/384.short

Kelley JH, Hanen MP (1988) Archaeology and the Methodology of Science. University of New Mexico Press, Albuquerque

King A (1999) Against structure: a critique of morphogenetic social theory. Socio Rev 47(2):199–227. URL http://onlinelibrary.wiley.com/doi/10.1111/1467-954X.00170/abstract

Kobti Z (2012) Simulating Household Exchange with Cultural Algorithms. In: Kohler TA, Varien MD (eds) Emergence and Collapse of Early Villages: Models of Central Mesa Verde Archaeology. University of California Press, Berkeley, pp 165–174

Kohler TA (2000) Putting Social Sciences Together Again: An Introduction to the Volume. In: Kohler TA, Gumerman GJ (eds) Dynamics in Human and Primate Societies: Agent-Based Modelling of Social and Spatial Processes. Santa Fe Institute Studies in the Sciences of Complexity. Oxford University Press, New York, pp 1–44

Kohler TA, Gumerman GJ (eds) (2000) Dynamics in Human and Primate Societies: Agent Based Modeling of Social and Spatial Processes. Oxford University Press, Oxford

Kohler TA, van der Leeuw SE (2007a) Introduction: Historical Socionatural Systems and Models. In: Kohler TA, van der Leeuw SE (eds) The Model-Based Archaeology of Socionatural Systems. School for Advanced Research Press, Santa Fe, pp 1–12

Kohler TA, van der Leeuw SE (eds) (2007b) The Model-Based Archaeology of Socionatural Systems. School for Advanced Research Press, Santa Fe

Kohler TA, Varien MD (2012) Emergence and Collapse of Early Villages in the Central Mesa Verde: An Introduction. In: Emergence and Collapse of Early Villages in the Central Mesa Verde: Models of Central Mesa Verde Archaeology. University of California Press, Berkeley, pp 1–14

Kohler TA, Kresl J, West CV, Carr E, Wilshusen RH (2000) Be There Then: A Modeling Approach to Settlement Determinants and Spatial Efficiency Among Late Ancestral Pueblo Populations of the Mesa Verde Region, U.S. Southwest. In: Kohler TA, Gumerman GJ (eds) Dynamics in Human and Primate Societies. Santa Fe Institute Studies in the Sciences of Complexity. Oxford University Press, New York, pp 145–178

Kohler TA, Gumerman GJ, Reynolds RG (2005) Simulating ancient societies. Sci Am 293:76–84

Kohler TA, Bocinsky RK, Cockburn D, Crabtree SA, Varien MD, Kolm KE, Smith S, Ortman SG, Kobti Z (2012a) Modelling prehispanic Pueblo societies in their ecosystems. Ecol Model 241:30–41. URL http://www.sciencedirect.com/science/article/pii/S0304380012000038

Kohler TA, Cockburn D, Hooper PL, Bocinsky RK, Kobti Z (2012b) The coevolution of group size and leadership: an agent-based public goods model for prehispanic pueblo societies. Adv Complex Syst 15(1 & 2):1150,007–1–1150,007–29. DOI 10.1142/S0219525911003256. URL http://www.worldscientific.com/doi/abs/10.1142/S0219525911003256

Kolm KE, Smith SM (2012) Modeling Paleohydrological System Strucure and Function. In: Kohler TA, Varien MD, Wright AM (eds) Emergence and Collapse of Early Villages: Models of Central Mesa Verde Archaeology. University of California Press, Berkeley, pp 73–83

Laird J (1919) The law of parsimony. Monist 29(3):321–344. URL http://www.jstor.org/stable/27900747

Lake MW (2000) MAGICAL Computer Simulation of Mesolithic Foraging. In: Kohler TA, Gumerman GJ (eds) Dynamics in Human and Primate Societies: Agent-Based Modelling of Social and Spatial Processes. Oxford University Press, New York, pp 107–143

Lake MW (2001a) Numerical Modelling in Archaeology. In: Brothwell DR, Pollard AM (eds) Handbook of Archaeological Sciences. Wiley, Chichester, pp 723–732

Lake MW (2001b) The use of pedestrian modelling in archaeology, with an example from the study of cultural learning. Environ Plann B Plann Des 28:385–403

Lake MW (2004) Being in a Simulacrum: Electronic Agency. In: Gardner A (ed) Agency Uncovered: Archaeological Perspectives on Social Agency, Power and Being Human. UCL Press, London, pp 191–209

Lake MW (2010) The Uncertain Future of Simulating the Past. In: Costopoulos A, Lake M (eds) Simulating Change: Archaeology into the Twenty-First Century. University of Utah Press, Salt Lake City, pp 12–20

Lake MW (2014) Trends in archaeological simulation. J Archaeol Method Theory. DOI 10.1007/s10816-013-9188-1, URL http://www.springerlink.com/openurl.asp?genre=article&id=doi:10.1007/s10816-013-9188-1

Lake MW, Crema ER (2012) The cultural evolution of adaptive-trait diversity when resources are uncertain and finite. Adv Complex Syst 15(1 & 2):1150,013–1–1150,013–19. DOI 10.1142/S0219525911003323, URL http://dx.doi.org/10.1142/S0219525911003323

van der Leeuw S, Redman CL (2002) Placing archaeology at the center of socio-natural studies. Am Antiq 67(4):597–605

van der Leeuw SE (2008) Climate and society: lessons from the past 10000 years. AMBIO: A J Hum Environ 37(sp14):476–482. URL http://www.bioone.org/doi/abs/10.1579/0044-7447-37.sp14.476

Levins R (1966) The strategy of model building in population biology. Am Sci 54(4):421–431. URL http://www.jstor.org/stable/10.2307/27836590

May RM (1976) Simple mathematical models with very complicated dynamics. Nature 261:459–467

McGlade J (1995) Archaeology and the ecodynamics of human-modified landscapes. Antiquity 69:113–132

McGlade J (1997) The Limits of Social Control: Coherence and Chaos in a Prestige-Goods Economy. In: van der Leeuw SE, McGlade J (eds) Time, Process and Structured Transformation in Archaeology. Routledge, London, pp 298–330

McGlade J (2005) Systems and Simulacra: Modeling, Simulation, and Archaeological Interpretation. In: Maschner HDG, Chippindale C (eds) Handbook of Archaeological Methods. Altamira Press, Oxford, pp 554–602

Mithen S, Reed M (2002) Stepping out: A computer simulation of hominid dispersal from Africa. J Hum Evol 43:433–462

Mithen SJ (1987) Modelling decision making and learning by low latitude hunter gatherers. Eur J Oper Res 30(3):240–242. URL http://www.sciencedirect.com/science/article/pii/0377221787900646

Mithen SJ (1988) Simulation as a Methodological Tool: Inferring Hunting Goals from Faunal Assemblages. In: Ruggles CLN, Rahtz SPQ (eds) Computer Applications and Quantitative Methods in Archaeology 1987, no. 393 in International Series, British Archaeological Reports, pp 119–137

Mithen SJ (1989) Modeling hunter-gatherer decision making: complementing optimal foraging theory. Hum Ecol 17:59–83. URL http://www.jstor.org/stable/4602911

Mithen SJ (1990) Thoughtful Foragers: A Study of Prehistoric Decision Making. Cambridge University Press, Cambridge

Mithen SJ (1991) 'A cybernetic wasteland'? Rationality, emotion and Mesolithic foraging. Proc Prehistoric Soc 57:9–14

Mithen SJ (1994) Simulating Prehistoric Hunter-Gatherers. In: Gilbert N, Doran J (eds) Simulating Societies: The Computer Simulation of Social Phenomena. UCL Press, London, pp 165–193

Nikitas P, Nikita E (2005) A study of hominin dispersal out of Africa using computer simulations. J Hum Evol 49:602–617

Oatley K, Johnson-Laird P (1987) Towards a cognitive theory of emotions. Cognit Emot 1:1–29

Odling-Smee FJ, Laland KN, Feldman MW (2003) Niche Construction: The Neglected Process in Evolution. Princeton University Press, Princeton, NJ

O'Meara T (1997) Causation and the struggle for a science of culture. Curr Anthropol 38(3):399–418. URL http://www.jstor.org/stable/10.1086/204625

Orton C (1982) Computer simulation experiments to assess the performance of measures of quantity of pottery. World Archaeol 14:1–19

O'Sullivan D, Haklay M (2000) Agent-based models and individualism: Is the world agent-based? Environ Plann A 32:1409–1425. URL http://www.envplan.com.libproxy.ucl.ac.uk/epa/abstracts/a32/a32140.html

Pickles J (1999) Arguments, Debates, and Dialogues: The GIS-Social Theory Debate and the Concern for Alternatives. In: Longley PA, Goodchild MF, Maguire DJ, Rhind DW (eds) Geographical Information Systems: Principles, Techniques, Applications, and Management. Wiley, New York, pp 49–60

Pinker S (2002) The Blank Slate: The Modern Denial of Human Nature. Viking, New York

Piou C, Berger U, Grimm V (2009) Proposing an information criterion for individual-based models developed in a pattern-oriented modelling framework. Ecol Modell 220(17):1957–1967. URL http://www.sciencedirect.com/science/article/pii/S030438000900324X

Powell A, Shennan S, Thomas MG (2009) Late Pleistocene demography and the appearance of modern human behavior. Science 324:1298–1301

Premo LS (2005) Patchiness and Prosociality: An Agent-Based Model of Plio/Pleistocene Hominid Food Sharing. In: Davidsson P, Takadama K, Logan B (eds) MABS 2004, Lecture Notes in Artificial Intelligence, vol 3415. Springer, Berlin, pp 210–224

Premo LS (2007) Exploratory Agent-Based Models: Towards an Experimental Ethnoarchaeology. In: Clark JT, Hagemeister EM (eds) Digital Discovery: Exploring New Frontiers in Human Heritage. CAA 2006. Computer Applications and Quantitative Methods in Archaeology, Archeolingua Press, Budapest, pp 29–36

Premo LS (2008) Exploring Behavioral Terra Incognita with Archaeological Agent-Based Models. In: Frischer B, Dakouri-Hild A (eds) Beyond Illustration: 2D and 3D Technologies as Tools of Discovery in Archaeology. British Archaeological Reports International Series. ArchaeoPress, Oxford, pp 46–138

Premo LS (2010) Equifinality and Explanation: The Role of Agent-Based Modeling in Postpositivist Archaeology. In: Costopoulos A, Lake M (eds) Simulating Change: Archaeology into the Twenty-First Century. University of Utah Press, Salt Lake City, pp 28–37

Premo LS (2012) Local extinctions, connectedness, and cultural evolution in structured populations. Adv Complex Syst 15(1&2):1150,002–1–1150,002–18. DOI 10.1142/S0219525911003268, URL http://www.worldscientific.com/doi/abs/10.1142/S0219525911003268

Premo LS, Scholnick JB (2011) The spatial scale of social learning affects cultural diversity. Am Antiq 76(1):163–176. URL http://saa.metapress.com/index/A661T246K0J1227K.pdf

Premo LS, Murphy JT, Scholnick JB, Gabler B, Beaver J (2005) Making a case for agent-based modeling. Soc Archaeol Sci Bull 28(3):11–13

Preston J (2014) Positivist and post-positivist philosophy of science. In: Oxford Handbook of Archaeological Theory. Oxford University Press, Oxford

Railsback SF, Grimm V (2012) Agent-Based and Individual-Based Modeling: A Practical Introduction. Princeton University Press, Princeton

Renfrew AC (1981) The Simulator as Demiurge. In: Sabloff JA (ed) Simulations in Archaeology. University of New Mexico Press, Albuquerque, pp 283–306

Renfrew C (1994) Towards a Cognitive Archaeology. In: Renfrew C, Zubrow EBW (eds) The Ancient Mind: Elements of a Cognitive Archaeology. Cambridge University Press, Cambridge, pp 3–12

Renfrew C, Poston T (1979) Discontinuities in the endogeneous change of settlement pattern. In: Transformations: Mathematical Approaches to Culture Change. Academic Press, New York, pp 437–461

Reynolds RG (1987) A production system model of hunter-gatherer resource scheduling adaptations. Eur J Oper Res 30(3):237–239. URL http://www.sciencedirect.com/science/article/pii/0377221787900634

Rubio-Campillo X, María Cela J, Hernàndez Cardona F (2011) Simulating archaeologists? Using agent-based modelling to improve battlefield excavations. J Archaeol Sci 39:347–356. URL http://www.sciencedirect.com/science/article/pii/S0305440311003475

Russell S (1999) Rationality and Intelligence. In: Woolridge M, Rao A (eds) Foundations of Rational Agency. Kluwer Academic, Dordrecht, pp 11–33

Russell S, Norvig P (2003) Artificial Intelligence: A Modern Approach, 2nd edn. Pearson Education, Upper Saddle River, NJ

Scaff LA (2011) Georg Simmel. In: The Wiley-Blackwell Companion to Major Social Theorists, vol 1. Blackwell, Chichester

Schuster HG (1988) Deterministic Chaos. VCH, New York

Searle J (1992) The Rediscovery of the Mind, chap Reductionsim and the Irreducibility of Consciousness. MIT Press, Cambridge, MA

Shanks M, Tilley C (1987a) Re-constructing Archaeology. University Press, Cambridge
Shanks M, Tilley C (1987b) Social Theory and Archaeology. Polity Press, Cambridge
Simon HA (1956) Rational choice and the structure of the environment. Psychol Rev 63(2):129–138
Slingerland E, Collard M (2012) Introduction. Creating Consilience: Toward a Second Wave. In: Slingerland E, Collard M (eds) Creating Consilience: Integrating the Sciences and Humanities. Oxford University Press, Oxford, pp 123–740 (e-edition)
Surovell T, Brantingham P (2007) A note on the use of temporal frequency distributions in studies of prehistoric demography. J Archaeol Sci 34(11):1868–1877. URL http://www.sciencedirect.com/science/article/pii/S030544030700012X
Thomas DH (1972) A Computer Simulation Model of Great Basin Shoshonean Subsistance and Settlement. In: Clarke DL (ed) Models in Archaeology. Methuen, London, pp 671–704
Thomas J (1988) Neolithic explanations revisited: The mesolithic–neolithic transition in Britain and south Scandinavia. Proc Prehistoric Soc 54:59–66
Thomas J (1991) The hollow men? a reply to Steven Mithen. Proc Prehistoric Soc 57:15–20
Thomas J (2004) Archaeology and Modernity. Routledge, London
Toffoli T, Margolus N (1987) Cellular Automata Machine: A New Environment for Modeling. MIT Press, Cambridge, MA
Waldrop M (1992) Complexity: The Emerging Science at the Edge of Order and Chaos. Simon & Schuster, New York
Watkins JWN (1952) Ideal types and historical explanation. Br J Philos Sci 3:22–43
Watson PJ, LeBlanc SA, Redman C (1971) Explanation in Archaeology: An Explicitly Scientific Approach. Columbia Univesity Press, New York
Wheatley D (1993) Going over Old Ground: GIS, Archaeological Theory and the Act of Perception. In: Andresen J, Madsen T, Scollar I (eds) Computing the Past: Computer Applications and Quantitative Methods in Archaeology 1992. Aarhus University Press, Aarhus, pp 133–138
Wilkinson T, Christiansen J, Ur J, Widell M, Altaweel M (2007) Urbanization within a dynamic environment: modeling Bronze Age communities in upper Mesopotamia. Am Anthropol 109(1):52–68. URL http://onlinelibrary.wiley.com/doi/10.1525/aa.2007.109.1.52/abstract
Wobst HM (1974) Boundary conditions for Palaeolithic social systems: A simulation approach. Am Antiq 39:147–178
Wobst HM (2010) Discussant's Comments, Computer Simulation Symposium, Society for American Archaeology. In: Costopoulos A, Lake M (eds) Simulating Change: Archaeology into the Twenty-First Century. University of Utah Press, Salt Lake City, pp 9–11
Wolfram S (1984) Cellular automata as models of complexity. Nature 311:419–424
Wright DJ, Goodchild MF, Proctor JD (1997) Demystifying the persistent ambiguity of GIS as "tool" versus "science". Ann Assoc Am Geogr 87:346–362. URL http://www.jstor.org/stable/2564374, also available from http://dusk.geo.orst.edu/annals.html (accessed 11/10/2004)
Wright HT, Zeder M (1977) The Simulation of a Linear Exchange System Under Equilibrium Conditions. In: Earle TK, Ericson JE (eds) Exchange Systems in Prehistory. Academic Press, New York, pp 233–253
Xue JZ, Costopoulos A, Guichard F (2011) Choosing fitness-enhancing innovations can be detrimental under fluctuating environments. PloS One 6(11):e26,770
Zeeman EC (ed) (1977) Readings in Catastrophe Theory. Addison-Wesley, Reading, MA
Zubrow E (1981) Simulation as a Heuristic Device in Archaeology. In: Sabloff JA (ed) Simulations in Archaeology. University of New Mexico Press, Albuquerque, pp 143–188

Chapter 2
Modeling Archaeology: Origins of the Artificial Anasazi Project and Beyond

Alan C. Swedlund, Lisa Sattenspiel, Amy L. Warren, and George J. Gumerman

2.1 Introduction

In 1994 the relatively young Santa Fe Institute (SFI) was moving to its new campus complex at the edge of the city of Santa Fe, New Mexico, USA. Researchers interested in the science of complexity were anxious to participate in this exciting new enterprise. A core resident faculty was being established, while at the same time numerous visiting scholars were invited to present new research and participate, for various lengths of time, in shaping SFI's identity and future trajectory. Within a year's time an important collaboration formed, somewhat by accident, which was to become known as the Artificial Anasazi Project.

Joshua Epstein and Robert Axtell, then of the Brookings Institution in Washington, DC, had come to present their agent-based modeling approach which they called "Sugarscape" (e.g. Epstein and Axtell 1996). In the course of their presentations they mentioned that they were interested in "real-world" applications of the model that could address questions about human behavior. In the audience happened to be George J. Gumerman, a southwestern archaeologist involved with SFI and its emerging program in cultural complexity. Gumerman, along with Jeffrey Dean and colleagues at the University of Arizona, had been collecting detailed

A.C. Swedlund (✉)
University of Massachusetts-Amherst, Amherst, MA 01003, USA
e-mail: swedlund@anthro.umass.edu

L. Sattenspiel • A.L. Warren
University of Missouri-Columbia, Columbia, MO 65211, USA
e-mail: sattenspiell@missouri.edu; alwgyb@mail.missouri.edu

G.J. Gumerman
Santa Fe Institute, 1399 Hyde Park Rd, Santa Fe, NM 87501, USA
e-mail: gjg@sarsf.org

© Springer International Publishing Switzerland 2015   37
G. Wurzer et al. (eds.), *Agent-based Modeling and Simulation in Archaeology*,
Advances in Geographic Information Science, DOI 10.1007/978-3-319-00008-4_2

environmental and archaeological data in the Kayenta Region of northeastern Arizona since the 1970s. They had constructed a huge database of high-quality environmental data as well as extensive information on temporal changes in climate and human habitation of the Long House Valley and other sites in the region. Although maize cultivation spread throughout the American Southwest over several centuries BC, the ancestral Pueblo Indians living in the sites of the Kayenta Region actively adopted agriculture commencing around AD 200. During the period of occupation between approximately AD 800 and AD 1300 the population grew significantly and then collapsed, and the region was abandoned between AD 1275 and AD 1300.

Axtell and Epstein had a powerful agent-based model, ready to be tested. Gumerman, Dean, and colleagues had an extraordinary dataset with a long timeline on which to test it. A research team was formed on the spot, and by 1996 preliminary runs were underway, with limited and often unsatisfying results. Steve McCarroll and Miles Parker were brought on as modeling assistants to Epstein and Axtell, and Alan Swedlund was invited to participate to provide demographic parameters that would be necessary to reflect human population dynamics. The intention was to compare the known settlement and population growth history of the Anasazi with the results generated using Ascape, an extension of the Sugarscape software. In Ascape agents occupy, grow, and populate a representational space designed to address specific research questions of interest. Resources (e.g. food, water, habitation sites, etc.) are distributed in the simulated space and that distribution can change over time if desired. Agents are given decision-making properties that allow them to move, garner resources, and interact with one another. Agents can also grow in numbers over time, necessitating movement as occupied areas become packed and food and other resources become scarce.

The space used in the Artificial Anasazi model is based on detailed paleoenvironmental, climatological, and archaeological data from the Long House Valley in northeastern Arizona.[1] Figure 2.1 shows the time series of several of these types of data. The data were used to design a landscape (analogous to Epstein and Axtell's Sugarscape) of annual variations in potential maize production values based on empirical reconstructions of low- and high-frequency paleoenvironmental variability in the valley (see Fig. 2.2a–c). The production values represent as closely as possible the actual production potential of various segments of the Long House Valley environment over the last 1,600 years. Historical settlements indicated in the archaeological record are also placed on the model landscape at their known locations within the valley (Fig. 2.2c).

For the initial Artificial Anasazi model it was decided that agents would be households instead of individuals, and households would be capable of clustering in areas where conditions permitted. Households would form much as human families do, averaging 5 individuals including male and female parents, and children of

---

[1]Although the environmental data are exceptionally fine-grained and detailed, they and the Artificial Anasazi model are not set within a continuous GIS spatial framework.

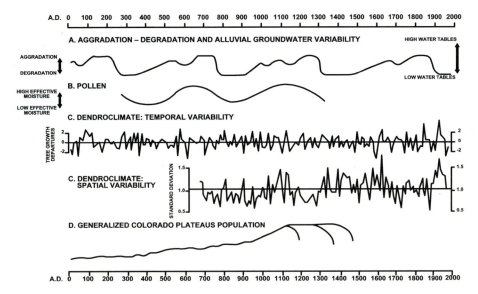

Fig. 2.1 Data used in the Artificial Anasazi model

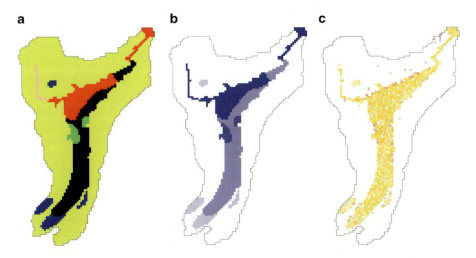

Fig. 2.2 (**a**) Environmental zones. (**b**) Hydrology. (**c**) Plot yields and historic settlements

varying age, although the activities of individual household members were not explicitly modeled. Each agent (household) was endowed with various attributes (e.g., life span, movement capabilities, nutritional requirements, consumption and storage capacities) in order to replicate important features of human households practicing horticulture. The limits, or rules, with which agents interacted with the environment, and with each other, were based on ethnographic reality and anthropological plausibility.

Agent (household) demographic patterns, subsistence, and movement behaviors were carefully built from the "bottom up." That is, deterministic mathematical models of growth or consumption were avoided and mortality, fertility, and consumption needs were based on a two-sex model of individuals of specific age. These were then summarized into households. In the model, the agents go through their life cycles on the empirically based landscape, adapting to changes in their physical environments. The agent-based simulations are then compared to the archaeological estimates generated empirically, and independently, by research archaeologists.

2.2   Structure of the Artificial Anasazi (AA) Model

The original Ascape implementation of the Artificial Anasazi (AA) model contains a number of versions designed for different purposes, in some cases for testing or batch runs and in other cases for experimentation with other types of model structures. We focus our initial discussion on the original AA model as described in Dean et al. (1998, 2000), Axtell et al. (2002), and Janssen (2009). Current efforts, described below, center on a version we call Artificial Long House Valley (ALHV). These two models differ in the basic demographic framework they use. The original AA model focuses on and depends solely on household-level information. "Births" relate to the origination of a new household, "deaths" refer to dissolution of an existing household through either death or abandonment. The ALHV model uses an individual-level framework, with each household's resident individuals considered explicitly.

The general structure of the AA model is shown in Fig. 2.3. The setup of the model is complicated and requires input of the detailed paleoenvironmental and historical data, construction of the valley, and construction and placement of households, settlements, and farm plots at appropriate locations in the valley, as well as determination of the initial harvest amounts available to each household.

Once the setup is complete, the model cycles through a series of steps that occur once a year for each household (Fig. 2.3). The first step is to determine whether each household has enough food available to satisfy its present needs. If there is enough food and the household is below a user-specified death age, then the amount of harvest resources available for the present year are estimated and added to the household's stores (assumed to last for 2 years). If food supplies are insufficient, the household has reached the specified maximum household age, or both, the household is abandoned and removed from the space. Following this series of decisions, households determine whether they have enough food for the coming year. If they do and they are within the user-specified fertility period, they can produce a new household through fissioning.[2] If a household does not have

---

[2]Note: although within the model this is specified through the naming of variables as reproduction, it is reproduction at the household, not individual level.

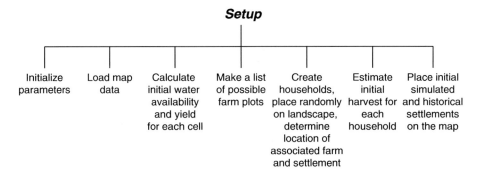

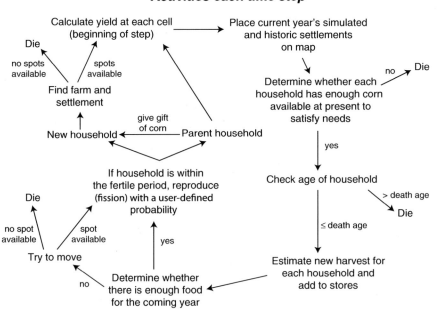

Fig. 2.3 The original Artificial Anasazi (AA) model

enough food in its stores for the coming year, it tries to move to a new location. This process involves a hierarchical series of decisions based on the distance from farm plots and water sources and the suitability of particular locations for the household. If a household is successful in finding a new location, it moves there and assesses whether fission is possible. If a location cannot be found, the household is abandoned and removed from the space. New households that originate after fissioning proceed through a process similar to the movement process of existing households to find suitable locations for their household and farm.

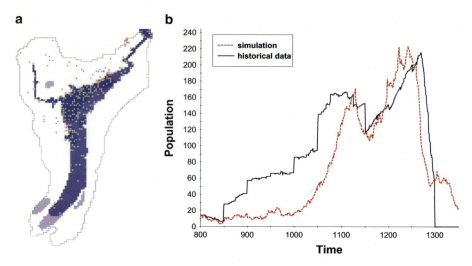

Fig. 2.4 Simulated (**a**) settlement distribution and (**b**) population vs. historical data, AA model

Once all households have completed the entire series of steps, the yield at each step is calculated and the cycle begins again. The simulation begins at AD 800 and runs yearly until AD 1350.

Summaries of outcomes of AA model simulations are given in Dean et al. (2000) and Axtell et al. (2002) (see also Janssen 2009). The earliest runs of the model often had the simulated population overshooting the archaeologically estimated population to a considerable degree, until it reached an apparent carrying capacity (e.g. see Dean et al. 2000, p.190–191; also Janssen 2009). Figure 2.4 provides an example of a simulation run that "fits" the historical data reasonably well. Analyses of the AA model indicate that simulation outcomes are quite variable and are also highly sensitive to the values chosen for the harvest variables (see also Janssen 2009). However, it is not yet clear whether the optimal values reflect more accurate estimates of real harvest potential than is possible using the unadjusted paleoclimatic data input into the model, or whether they are simply a result of adjustments to produce better curve fitting. This assessment requires careful attention to assumptions about how the variables are incorporated into the model as well as more extensive sensitivity analyses of how variation in their values influences model outcomes.

2.3 Selected Socio-Ecological Models Similar to Artificial Anasazi

Prior to the development of Artificial Anasazi, several efforts were aimed at simulating social change in the southwestern United States (Cordell 1972, Dove 1984 for example) but many of these models applied a top-down approach to simulation and hard-coded variables and interaction rules (Gumerman and Kohler 1996). Around the time the Artificial Anasazi research team was coming together, other agent-based models aimed at exploring social and ecological interactions with a bottom-up approach were also in development. Table 2.1 provides a description of several of the major archaeological agent-based models that have been developed. In this section we discuss a selection of those models that most closely relate in purpose and general structure to the Artificial Anasazi model.

The Evolution of Organized Society (EOS) model generated group-level behaviors such as information exchange, group decision-making, and the emergence of hierarchical social structures among foragers in Upper Paleolithic France by modeling the environment and resource-gathering behaviors (Doran et al. 1994). Another model, MAGICAL (Multi-Agent Geographically Informed Computer AnaLysis) (Lake 2000), was developed by combining geographic information systems software with multi-agent simulations to explore possible explanations for the distribution of flint artifacts in the Southern Hebrides. The incorporation of detailed GIS data into this model reinforced a trend in agent-based modeling toward the inclusion of high quality environmental data in models focused on socio-ecological changes.

Just as Artificial Anasazi was influenced by other simulations of social-ecological processes in the Southwest and in other areas, the success of Artificial Anasazi and similar models provided further support for the suitability of agent-based models for exploring changing human-environment interactions in a given geographic space and encouraged the continued development of other models. Most notable among these modeling efforts are the models developed by Kohler and colleagues that comprise the Village Ecodynamics Project (VEP), which commenced development around the same time as Artificial Anasazi and also came out of work being done at the Santa Fe Institute. The goals of this project include exploring the co-evolution of society and environment by accurately recreating the landscape, understanding the factors that contribute to complicated behaviors on this landscape, and developing an understanding of the factors that may have driven village aggregation, growth, and depopulation in southwestern Colorado between AD 600 and AD 1300 (see Kohler et al. 2007, 2012). Much like Artificial Anasazi, the VEP model generates maize production data based on climate, soil quality, and plots farmed, but it also incorporates over-farming as a factor (Kohler et al. 2007). In addition to the maize-related factors that largely drive the Artificial Anasazi simulations, the VEP model also incorporates social and cultural learning, water usage, wood use for fuel, and hunting parameters to explore resource procurement strategies and how such strategies may have contributed to demographic patterns observed in the archaeological record for the area, most notably the depopulation during the late thirteenth century AD (Kohler et al. 2007).

In a model similar to Artificial Anasazi and VEP, Griffin and Stanish (2007) have modeled the Lake Titicaca basin with the goal of exploring the role of environmental and social factors on the development of complex societies in the region. Their model specifically examined factors that led to this political consolidation, such as agriculture, migration, competition, and trade (Griffin and Stanish 2007). Like Artificial Anasazi, their model was situated during a specific

Table 2.1 Survey of similar agent-based models

Start of Project or Earliest Known Publication[a]	Who	Model Name (if available)	Focus
1994	Doran, Palmer, Gilbert, Mellars	Evolution of Organized Society (EOS)	Models Upper Paleolithic social change by focusing on human interactions within a shared environment and explores the factors that contribute to group formation
1996	Dean, Gumerman, Axtell, Epstein, Swedlund	Artificial Anasazi	Explores reasons for why the prehistoric Pueblo (Anasazi) people abandoned Long House Valley in Arizona by modeling how demographic and social constraints interacted with environmental factors in the time leading up to the abandonment of the valley
1999	DiPiazza, Pearthree		Explores theories about the expansion of the Lapita cultural complex across the Southern Pacific Ocean by modeling demography and migration patterns
2000	Lake	MAGICAL	Designed to integrate GIS and agent-based simulations and to form a basis from which other archaeologists could build their own models; the first application explored whether modeling small group foraging patterns could explain artifact distributions in the Southern Hebrides
2002	Kohler, Johnson, Varien, Ortman, Reynolds, and others	Village Ecodynamics Project	Models environmental and social interactions on an accurately recreated landscape to explore factors that drive village aggregation, growth, and depopulation in southwestern Colorado between AD 600 and AD 1300
2003	Brantingham		A neutral model that indicates that raw stone material patterning in the archaeological record matches procurement strategy that does not require adaptive optimization, planning or risk minimization

Year	Name	Researchers	Description
2007		Griffin, Stanish	Examines the social and ecological dynamics, including agriculture, migration, and trade, that led to patterns of consolidation and emergence of complex society in the Lake Titicaca Basin in South America
2007	Enkimdu	Wilkinson, Gibson, Christiansen, Widell, Schloen, and others	Explores the social and ecological dynamics that led to population change in Bronze Age Mesopotamia especially in times of economic or disease stress
2008		Conolly, Colledge, Shennan	Models the effects of drift and vertical transmission of cultural adaptations to explain the reasons for the reduction in crop variability during the European Neolithic
2008		Kowarik, Wurzer, Reschreiter, Rausch, Totschnig	Simulates the working processes of Bronze Age mines at Hallstatt to gain insight into factors that affect mine output and contribute to mine exhaustion
2009	Patron World	Graham	Models individual-level interactions including grievances and gift giving to explore the emergence of violence and civil unrest in ancient Rome
2010		Premo, Kuhn	Models the effect of local extinctions on diversity, differentiation between groups, and rates of cultural change with regard to stone tool evolution in the Paleolithic
2012		Campillo, Cela, Hernàndez Cardona	Simulates battles, projectile trajectory, site degradation, fieldwork techniques to determine most effective strategies for doing battleground fieldwork

[a]The earliest publications we could find for each of these projects are: Doran et al. (1994), Dean et al. (1998), DiPiazza and Pearthree (1999), Lake (2000), Kohler et al. (2007), Brantingham (2003), Griffin and Stanish (2007), Wilkinson et al. (2007), Conolly et al. (2008), Kowarik et al. (2009), Graham (2009), Premo and Kuhn (2010), Campillo et al. (2012)

time period in a specific region and aimed to increase understanding of social processes occurring in pre-state level agricultural societies. The Titicaca Basin model differed from Artificial Anasazi in that the ecological and agricultural factors shaping the environment were not incorporated with as fine a scale as in Artificial Anasazi and were instead included with the goal of simply creating a reasonable landscape in which the agents could operate. Additionally, this model was validated using multi-dimensional measures in order to fully account for the multiple factors included in the model that contributed to long-term political changes. Griffin and Stanish's Lake Titicaca Basin model continued the trend of modeling with realistic and empirically informed environments and behaviors that began with models like Artificial Anasazi, but also borrowed from a political science model designed to simulate nation-state long-term political change.

Another model, Enkimdu, was developed to look at social and ecological dynamics of population change in Bronze Age Mesopotamia (Wilkinson et al. 2007). Like Artificial Anasazi, this model incorporates natural processes such as weather, crop growth, hydrology, soil quality, and population dynamics, but it also models how social behaviors such as farming and herding practices, kinship-driven behaviors, and trade interact with these natural processes to cause concurrent, dynamic changes in both types of processes (Wilkinson et al. 2007). Relying on modeling concepts from other fields such as the Dynamic Information Architecture System (DIAS) and the Framework for Addressing Cooperative Extended Transactions (FACET), Wilkinson and colleagues explored patterns related to demography, subsistence, kinship, and reciprocal exchange in situations of social stress to test hypotheses related to changing social and ecological dynamics (Wilkinson et al. 2007).

The combination of ecological, agricultural, cultural, and political factors into one model as demonstrated in all these studies illustrates how interdisciplinary work can better inform archaeological agent-based models. These models are also especially valuable because they demonstrate how micro-level behaviors of agents, determined archaeologically and ethnographically, within a specific temporal and geographic space can result in the emergence of macro-level patterns that can also be compared with the archaeological record—an advantage of agent-based models over other forms of modeling (Griffin and Stanish 2007; Schelling 1978; Epstein 1999).

2.4 Current Efforts: The ALHV Model

Recent efforts working with the Artificial Anasazi project center on the ALHV model. This model differs from the AA model in only a few characteristics, but these differences have far-reaching implications for model structure and simulation results. As mentioned above, the primary difference is that individuals within households are now considered to be active agents. Rather than basing fertility and

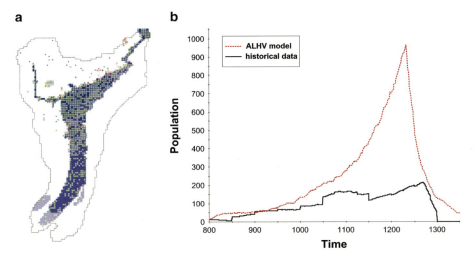

Fig. 2.5 Simulated (**a**) settlement distribution and (**b**) population vs. historical data, ALHV model

mortality levels on aggregate household-level probabilities as in the AA model, in the ALHV model individual agents are endowed with age-specific fertility and mortality schedules that govern their birth and death.

We are still in the process of determining the nature and impact of the specific differences between the two models, but household-level fission and death still do appear to be present. We have determined that they work somewhat differently in the ALHV model, but the particular mechanisms are not yet understood. Eventually the ALHV model, with its disaggregated, individual-level demographic processes, will be used for questions regarding the potential impacts on population growth of morbidity, mortality, and reduced fertility due to disease and food scarcity.

Figure 2.5 demonstrates that results of simulations of the AHLV model are dramatically different from those of the AA model (as shown in Fig. 2.4). The vast majority of runs exhibit a pattern of exponential growth (consistent with the assumption of constant age-specific fertility and mortality rates), followed by a steep decline as the region becomes agriculturally unproductive. Sensitivity analyses using various estimates for fertility, mortality, and harvest variables indicate that, as would be expected, the rate of exponential growth is sensitive to changes in the age-specific fertility and mortality parameters. Interestingly, however, unlike the AA model, the ALHV model outcomes are not influenced by changes in the values of harvest parameters. Clearly, in bringing the individual-level fertility and mortality processes into the model, household-level population control processes related to the environment and harvest potential were decoupled from the model. This decoupling has, however, removed any semblance of a "fit" between the simulations and the archaeological data, with the result that, at present, the ALHV model is not an adequate representation of the population history of the Long House Valley. Reincorporation of the environmental constraints, at a minimum, is required in order for the ALHV model to be a successful model for the rise and abandonment of the Long House Valley.

More importantly, influences such as disease, nutrition, migration, and warfare operate most strongly at the individual level, and these processes have been proposed repeatedly in explanations of the population history of the region. In order for models to pursue questions related to these processes in the Long House Valley and elsewhere, both individual-level demography and household-level environmental constraints are essential components. Thus, present research on the ALHV model is directed at solving the problem of recoupling and integrating individual-level and environmental constraints on population growth.

2.5   Discussion and Conclusions

Artificial Anasazi, its predecessors, and its successors have demonstrated the usefulness of agent-based modeling in archaeological applications. Because agent-based modeling in archaeology allows for hypothesis testing, verifying the accuracy of other methodologies, formalization of theories, exploration, and experimentation that are not always possible with other lines of archaeological inquiry, the continued application of agent-based modeling methodologies to archaeological questions is essential (Kowarik 2011).

In our own work, we envision the following as new directions for Artificial Anasazi and Artificial Long House Valley:

- More extensive and systematic sensitivity analyses
- Incorporation of explicit infectious disease processes, nutritional stresses, violence
- Interactions between the Long House Valley and Black Mesa, a major archaeological region adjacent to the Long House Valley

We also suggest that future applications of ABMs to archaeological questions should consider the following:

- Continued focus on modeling individual-level interactions
- More focus on emergent social patterns that result from local behaviors and socio-ecological dynamics and comparison of these simulated patterns to patterning observed in the archaeological record
- Continued interdisciplinary work

We see the most important goal of our work in the near term, once the ALHV model is up and running well, to be the development of insights into the varying roles of disease, diet, and fertility in the decline and eventual abandonment of the Long House Valley. Once we understand the likely possibilities for the impact of these demographic and nutritional constraints, we may gain clearer understandings of the influence of political and social factors in triggering final abandonment of the region.

Acknowledgements Jeff Dean, Rob Axtell, Josh Epstein, and Miles Parker were key players in the original development of the Artificial Anasazi model. This paper and the ongoing work with the model would not be possible without their contributions.This work was made possible by support of the Santa Fe Institute and by the two senior authors' attendance as Short Term Visitors at the National Institute for Mathematical and Biological Synthesis (NIMBioS), an Institute sponsored by the National Science Foundation, the US Department of Homeland Security, and the US Department of Agriculture through NSF Awards #EF-0832858 and #DBI-1300426, with additional support from The University of Tennessee, Knoxville. Any opinions, findings, and conclusions or recommendations expressed in this paper are those of the authors and do not necessarily reflect the views of the National Science Foundation.

References

Axtell RL, Epstein JM, Dean JS, Gumerman GJ, Swedlund AC, Harburger J, Chakravarty S, Hammond R, Parker J, Parker M (2002) Population growth and collapse in a multiagent model of the Kayenta Anasazi in Long House Valley. Proc Natl Acad Sci USA 99(3):7275–7279. DOI 10.1073/pnas.092080799

Brantingham PJ (2003) A neutral model of stone raw material procurement. Am Antiq 68(3): 487–509

Campillo XR, Cela JM, Hernàndez Cardona FX (2012) Simulating archaeologists? Using agent-based modelling to improve battlefield excavations. J Archaeol Sci 39(2):347–356. DOI http://dx.doi.org/10.1016/j.jas.2011.09.020, URL http://www.sciencedirect.com/science/article/pii/S0305440311003475

Conolly J, Colledge S, Shennan S (2008) Founder effect, drift, and adaptive change in domestic crop use in early Neolithic Europe. J Archaeol Sci 35(10):2797–2804. DOI http://dx.doi.org/10.1016/j.jas.2008.05.006, URL http://www.sciencedirect.com/science/article/pii/S030544030800112X

Cordell L (1972) Settlement Pattern Changes at Wetherill Mesa, Colorado: A Test Case for Computer Simulations in Archaeology. PhD thesis, University of California, Santa Barbara

Dean JS, Gumerman GJ, Epstein JM, Axtell RL, Swedlund AC, Parker MT, McCarroll S (1998) Understanding Anasazi Culture Change Through Agent-Based Modeling. In: Kohler TA, Gumerman GJ (eds) Working Paper No.98-10-094, Santa Fe Institute Studies in the Sciences of Complexity, Santa Fe Institute, Santa Fe, pp 179–205

Dean JS, Gumerman GJ, Epstein JM, Axtell RL, Swedlund AC, Parker MT, McCarroll S (2000) Understanding Anasazi Culture Change Through Agent-Based Modeling. In: Kohler TA, Gumerman GJ (eds) Dynamics in Human and Primate Societies: Agent-Based Modelling of Social and Spatial Processes. Santa Fe Institute Studies in the Sciences of Complexity. Oxford Univesity Press, New York, pp 179–205

DiPiazza A, Pearthree E (1999) The spread of the 'Lapita people': a demographic simulation. J Artif Soc Soc Simulat 2(3):1–15

Doran JE, Palmer M, Gilbert N, Mellars P (1994) The EOS Project: Modelling Upper Palaeolithic Social Change. In: Gilbert N, Doran J (eds) Simulating Societies: The Computer Simulation of Social Phenomena. UCL Press, London, pp 195–221

Dove D (1984) Prehistoric Subsistence and Population Change Along the Lower Agua Fria River, Arizona: A Model Simulation. In: Anthropological Research Papers 32. Arizona State University, Tempe

Epstein JM (1999) Agent-based computational models and generative social science. Generative Soc Sci Stud Agent Based Comput Model 4(5):41–60

Epstein JM, Axtell R (1996) Growing Artificial Societies: Social Science from the Bottom Up. MIT Press, Cambridge, MA

Graham S (2009) Behaviour space: Simulating Roman social life and civil violence. Digital Stud/Le Champ Numérique 1(2). URL http://www.digitalstudies.org/ojs/index.php/digital_studies/article/view/172

Griffin AF, Stanish C (2007) An agent-based model of prehistoric settlement patterns and political consolidation in the Lake Titicaca Basin of Peru and Bolivia. Struct Dynam 2(2), URL http://www.escholarship.org/uc/item/2zd1t887

Gumerman GJ, Kohler TA (1996) Creating Alternative Cultural Histories in the Prehistoric Southwest: Agent-Based Modeling in Archaeology (a Progress Report). In: Examining the Course of Southwest Archaeology: The Durango Conference, September 15, 1995, New Mexico Archaeological Council, Durango, CO. URL http://www.santafe.edu/media/workingpapers/96-03-007.pdf

Janssen MA (2009) Understanding Artificial Anasazi. J Artif Soc Soc Simulat 12(4):13. URL http://jasss.soc.surrey.ac.uk/12/4/13.html

Kohler TA, Johnson CD, Varien M, Ortman S, Reynolds R, Kobti Z, Cowan J, Kolm K, Smith S, Yap L (2007) Settlement Ecodynamics in the Prehispanic Central Mesa Verde Region. In: Kohler TA, van der Leeuw S (eds) The Model-Based Archaeology of Socionatural Systems. School for Advanced Research Press, Santa Fe, pp 61–104

Kohler TA, Bocinsky RK, Cockburn D, Crabtree SA, Varien MD, Kolm KE, Smith S, Ortman SG, Kobti Z (2012) Modelling prehispanic Pueblo societies in their ecosystems. Ecol Model 241:30–41. URL http://www.sciencedirect.com/science/article/pii/S0304380012000038

Kowarik K (2011) Agent-Based Modelling and Archaeology. Paper Presented at the First Trans-Disciplinary Workshop on Agents in Archeology, Natural History Museum Vienna, Austria. URL http://aia11.nhm-wien.ac.at/downloads/AIA11-Kowarik-ABM-in-Archeology.pdf

Kowarik K, Wurzer G, Reschreiter H, Rausch A, Totschnig R (2009) Mining with Agents. Agent-Based Modeling for the Bronze Age Saltmine of Hallstatt. In: Archäologie und Computer Workshop 13, Museen der Stadt Wien

Lake MW (2000) Computer Simulation of Mesolithic Foraging. In: Gumerman G, Kohler T (eds) Dynamics in Human and Primate Societies: Agent-Based Modeling of Social and Spatial Processes. Oxford University Press, New York, pp 107–143

Premo LS, Kuhn SL (2010) Modeling effects of local extinctions on culture change and diversity in the paleolithic. PLoS One 5(12):e15,582. DOI 10.1371/journal.pone.0015582, URL http://dx.doi.org/10.1371%2Fjournal.pone.0015582

Schelling TC (1978) Micromotives and Macrobehavior. W. W. Norton & Company, New York

Wilkinson TJ, Gibson M, Christiansen JH, Widell M, Schloen D, Kouchoukos N, Woods C, Sanders J, Simunich KL, Altaweel M, Ur JA, Hritz C, Lauinger J, Paulette T, Tenney J (2007) Modeling Settlement Systems in a Dynamic Environment: Case Studies from Mesopotamia. In: Kohler TA, van der Leeuw SE (eds) The Model Based Archaeology of Socionatural Systems. School for Advanced Research Press, Santa Fe, New Mexico, pp 175–208

Part II
Methods

Chapter 3
Agent-Based Simulation in Archaeology: A Characterization

Felix Breitenecker, Martin Bicher, and Gabriel Wurzer

3.1 Motivation and Classification of Computer Models

The phrase "dare to know" (Latin "*sapere aude*", originally brought forward by the Roman poet Horace and later made popular by Kant (1784)) is famously known as the motto of the enlightenment, representative for the intellectual attitude of the western civilisation during seventeenth and eighteenth century. From this era also come some of the core concepts used in simulation, *experiment* and *observation*: An experiment is a procedure that seeks to prove or reject a hypothesis (Kant 1787, p. 23) or observe an emergent outcome. However, there are cases in which questions cannot be answered by conducting classical experiments, because

- these would be too expensive (e.g. investigate the earthquake resistance of a skyscraper)
- these would be too dangerous or not possible for ethical reasons (e.g. investigating the spread of an epidemic)
- these would not be possible for logical reasons (e.g. measuring the kernel temperature of the sun)
- these would not be possible, as information from the future is required (e.g. meteorological weather prognosis)
- these would not be possible, as information from the past is required (e.g. proving or falsifying historical hypotheses if archaeological findings are missing)

Therefore, instead of conducting "real" experiments, researchers have turned to *experimentation in silico*, in which *simulation* plays the role of a tool (or method,

F. Breitenecker (✉) • M. Bicher • G. Wurzer
Vienna University of Technology, Karlsplatz 13, 1040 Vienna, Austria
e-mail: felix.breitenecker@tuwien.ac.at; martin.bicher@tuwien.ac.at; gabriel.wurzer@tuwien.ac.at

© Springer International Publishing Switzerland 2015   53
G. Wurzer et al. (eds.), *Agent-based Modeling and Simulation in Archaeology*,
Advances in Geographic Information Science, DOI 10.1007/978-3-319-00008-4__3

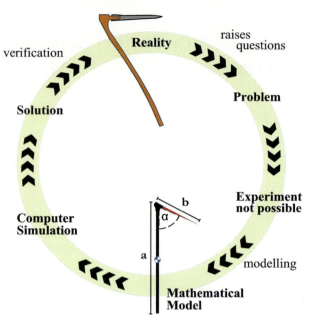

Fig. 3.1 Motivation and life cycle of mathematical modelling

see Fig. 3.1): A real entity raises questions, which leads to a problem that is either defined explicitly or (as it often is in archeology) informally. The problem cannot be answered using classical experimentation, and therefore, a mathematical or theoretical model is built, which imitates the behaviour and structure of the real system as closely as possible. Based on this model, a simulation is run to solve the given problem, supported by initial data in order to receive "satisfying" results. The level of detail and precision within the modelling process and the amount and quality of the initial data (consisting of input and initial conditions of the simulation) are mainly responsible for the validity of the simulation results and thus the reliability of the answer to the given problem. However, this does not free the researcher of a verification phase, which examines how congruent the model is in respect to reality. A closer look at this aspect is given in Chap. 2.

3.1.1   The Model: Creating an Abstraction of Reality

The word "model" derives from the Latin word "modulus", which was used as a measure in architecture until the eighteenth century. To be more precise, "modulus" was the base unit used in architectural planning on which plans and miniature versions of buildings were based. Its meaning slowly transitioned into the now-common "model", which signifies a simplified replica of the real world. With respect to Stachowiak (1973), a model is characterized by at least three different features (translated from German):

1. Mapping—A model is a representation of a natural or an artificial object.
2. Reduction—A model is simplified and does not have all attributes of the original object.
3. Pragmatism—A model is always created for a certain purpose, a certain subject and a certain time-span.

Similarly to models made out of real material (wood, stone, paper,...), mathematical models used for simulation also fulfil these three characteristics: On the one hand, a detailed mapping of the real world guarantees reliable results. On the other hand, it is still necessary to do simplifications in order to be able to perform the simulation at all. Although processing power and memory resources of computers are increasing rapidly, the ability to calculate simulations in-depth is still limited (also see high-performance computing, Chap. 4): One must not forget that the more complex the model gets, the more data is required to calibrate the simulation correctly. Regarding reduction, the following techniques can be used to obtain reasonable models:

- Distinction: Neglect those parts of reality which are not relevant for the purpose of the model.
- Cutback: Neglect details that are irrelevant.
- Decomposition: Decouple segments of an object which are not or weakly linked and treat them as own models.
- Aggregation: Couple segments which act alike and threat them as one.
- Abstraction: Create technological terms, classes and clusters in order to simplify the overall problem.

Surprisingly, the history of mathematical modelling is older than the history of the computer. For example the famous Lotka-Volterra Differential equations, a mathematical model formerly designed to simulate the cohabitation of a predator and a prey species, was initially developed by Alfred J. Lotka and Vito Volterra between 1925 and 1926 (Lotka 1956). Compared to that, the first (useful) computers were developed during the Second World War. Surely, the two scientists had problems to simulate their own derived differential equation model by hand, but they had some success regarding the qualitative analysis of their results. So, the idea of simulation is basically not linked to computers, although most simulations can only be executed with their support. Even after the introduction of computers, some simulations such as the evacuation model by Predtetschenski and Milinski (1971) were based on hand calculation methods, and some are even used to this day (Rogsch et al. 2008).

One of the most common definitions of the modern meaning of the word simulation is finally given by Shannon (1975): "Simulation is the process of designing a model of a real system and conducting experiments with this model for the purpose either of understanding the behaviour of the system and its underlying causes or of evaluating various designs of an artificial system or strategies for the operation of the system" Shannon (1975, p. 2).

Shannon defines the modelling process as a sub-process of the simulation, as the purpose of the model is solely given by the purpose of the simulation. On the one hand, it can be seen easily that the word computer is not part of this definition. On the other hand, since the invention of the computer the fields of application for mathematical models is steadily increasing as due to growing computational resources, a growing number of complex simulations can be executed.

Agent-based modelling is just one of many modelling approaches. In order to fully understand the basic mechanics of this specific approach, some specific terms related to modelling and simulation have to be discussed in the forthcoming sections. Very often, the explanation of these terms is supported by example models. These examples were only developed theoretically (i.e. for didactical reasons) and their validity and value must thus not been overestimated.

3.1.2   Black-Box and White-Box Models

Within system theory, the word *black-box* describes a basic input-output system in which the correlation between input and output is not known. Such an "opaque box" somehow translates the input signal into some output and does not reveal how the translation takes place. Regarding modelling and simulation, the term black-box is used if modelling approaches are mainly based on observations and explanations but not on strict laws and proven formulas. The model has to be developed only knowing about input data, a little bit of information about the behaviour of the real system and some output data required for the calibration of the model. If done properly, those kinds of models are also working reliably, even though everything in between the input and the output part can hardly be explained in a reasonable way; nevertheless, there are some authors who seek to reverse-engineer black-box models, in order to give users confidence and also some control over the work they are doing with them (Rogsch and Klingsch 2011).

White-box modelling, i.e. modelling based on proven laws and axioms, is preferred to black-box modelling. When this is not possible, a mixture between the approaches has to be used. The gradual transition between white-box and black-box techniques can be seen in Fig. 3.2b: Most problems that can be described by proven laws or axioms can be simulated using white-box models, whereas problems that are based on expert knowledge and observations will need to be approached by black-box modelling:

- *How fast does the pendulum swing?*
  This problem can be solved using physical basics like Newton's law of gravitation, friction formulas and the law of conservation of energy, resulting in the so called pendulum-equation. This second order differential equation can be solved by most of the now-common numerical computing and simulation packages. The derivation of the model was solely based on physical laws, which makes it a *White Box model*.

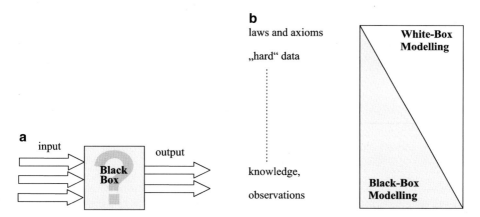

Fig. 3.2 (**a**) Black box model. (**b**) Fields of application for black box and white box models

- *How fast does a population grow?*
  Before being able to solve that problem, a lot of data has to be procured and investigated. Aside from the data, it can be observe that the growth rate of the population is proportional to the number of individuals (the higher the number of fertile individuals, the more individuals are born). Furthermore, the size of the population is limited due to resources and lack of space. In order to model these constraints governing the size of the population, a so-called logistic differential equation can be applied. The modelling process is basically motivated by data and observations, and thus is an example of a *Black Box model*.

Summing up, both examples result in mathematical formulae. The main difference being the derivation of the equation: In the first case, the formula is solely based on already-existing, proven laws. In the second case, the equation is gained by observation of the behaviour and data of a real system. Agent-based models are mostly used as black-box models, but are not limited to this field of application. Before we continue to the description of basic concepts used in agent-based modelling, some principles of microscopic and macroscopic modelling have to be discussed.

3.1.3 Microscopic and Macroscopic Approach

The larger and more complex the investigated system, the higher the need for a computer simulation (instead of a classical solution to a problem). Usually, very big complex systems consist of (or can at least be partitioned into) a large number of less complex subsystems which can be investigated in isolation. The final complexity of the overall system is governed by the interaction of all subsystems (also called *components* of the system). If those subsystems share the same basic properties and act similarly, the system can be described by an agent-based model. Examples for

Table 3.1 Examples of
systems made out of similar
components

System	Components
Population	Individuals
Company	Subsidiaries
Traffic	Cars
Paper	Wooden fibres

such systems and their components are given in Table 3.1. Interestingly, the scale
at which these systems are modelled may differ, as we may distinguish models as
being either *microscopic* or *macroscopic*:

- *Microscopic models* decompose a system into its subsystems. Each of these
  subsystems is modelled separately, taking into consideration that they are allowed
  to communicate with each other, in order to model the structure of the whole
  system. Because the subsystems are similar, they are usually modelled in the
  same way, but then supplied with different parameters. In order to decouple the
  system, simplifications have to be done regarding the often complex relationships
  between the subsystems. These rich relationships also make "the whole [...]
  greater than the sum of its parts" (Aristoteles), a property that is often referred to
  as *emergence*.
- In *Macroscopic models*, the influence of the subsystems upon the whole system
  is investigated and generalized, in order to be able to model the whole system
  at once. Often, black-box approaches have to be used in that context, leading to
  so-called "behavioural models" (i.e. models developed by studying the behaviour
  and causal relationships of the system). Although these models do not take into
  account that the whole system consists of subsystems, very satisfying results can
  be obtained if the model is developed properly.

The choice of the modelling approach is often a result of a "top-down" or
"bottom-up" philosophy being chosen for the abstraction of the model. Usually,
a "bottom-up" approach implies a microscopic model, in which the lowest level of
the system (i.e. the "individuals" having the highest degree of detail) is analyzed
first. After building up such a microscopic model, it is not unusual that the modeller
has to simplify it, as it would otherwise become too complex. Vice versa, a "top-
down" approach usually leads to a macroscopic model. The abstraction process then
starts at the top level (looking at the whole system), and iteratively goes into more
detail until the granularity is sufficient for the stated problem.

In order to further illustrate the difference between microscopic and macroscopic
models, let us assume an example where the problem is to simulate the degrees
of a large number of tourists visiting a museum after the occurrence of a fire.
The microscopic approach would regard the whole system (museum) through its
constituent parts (the visitors). These would be simulated individually insofar as
each simulated person is free to choose his own exit route. The interaction between
the individuals would here mean the movements, collisions and avoidance strategies
that the visitors follow (also referred to as *pedestrian dynamics*), leading to emergent
outcomes such as lane formation, congestive bulks before exits and collective

motion among the crowd (Helbing and Molnár 1995; Helbing et al. 2000). A macroscopic model would identify the mass of people to behave like a liquid, tending towards the areas with the least pressure (which in our case would be defined to be the exits). In this case, one would use the so-called Navier-Stokes-equations, which are used within physics for modelling fluid dynamics, to arrive at a macroscopic model for the system.

Summarizing, the decision which of the two mentioned approaches is taken requires a more detailed specification of the modelling purpose. It also depends on the number and quality of provided measurements and data that is necessary for calibrating the model. Both modelling approaches have advantages and disadvantages, as summarized in Table 3.2.

As computational resources are increasing exponentially (Moore 1965), microscopic models are getting more and more popular. Agent-based models are probably the best-known example of microscopic approaches, but not the only one.[1] A recent trend also lies in *cohort simulations*, which is a simulation technique in which individuals sharing the same properties are clustered into "cohorts". This approach lies in between the macroscopic and microscopic level and is therefore called *mesoscopic*.

Table 3.2 Advantages and disadvantages of microscopic and macroscopic models

Property	Microscopic	Macroscopic
Processing time	Usually microscopic models require lots of computational resources. So calculations are usually rather *slow*	As not all sub-models have to be evaluated separately computation is rather *fast*
Abstraction level, i.e. how difficult the explanation of the model is to non-experts	As the connection between model and reality is very obvious the necessary level of abstraction is relatively *low*	As the model is very often either developed black-box style or derived out of complex formulas the connection between model and reality is more difficult to understand. So the abstraction level is usually *high*
Number of necessary parameters	Due to the high number of sub-models *a lot* of specific parameters have to be defined which poses a difficult task for calibration of the model	Usually the model is well defined by *a few* parameters dealing with the overall dynamics of the system. Regarding commonly gained statistical data calibration is often easier too
Flexibility	Small changes within the structure of the sub-models can be performed very easily without restructuring the whole model. Thus microscopic models are *very flexible*	Small structural changes usually require redefinition of the whole model (unless it can- be corrected by parameter variations). So macroscopic models are *hardly flexible*

---

[1]Others types would for example be cellular automata (see e.g. Wolfram 2002, p.8) or finite elements.

3.2   Introduction to Agent-Based Modelling

The term "agent" derives from the latin word "agere", meaning "to act". So, defined by the root of the word itself, the main function of an agent is to act—individually. Hence, we can give a first characterization: *Agent-based simulation is micro-simulation based on individually (inter)acting sub-models.* One aspect of this sentence is especially noteworthy: It does not state that agents necessarily have to pose for persons. Rather, a sub-model can represent everything ranging from a car in a traffic jam to a charged molecule in a liquid or even an abstract concept that can be represented as an agent. Regardless of what is being represented by an agent, there are still some additional requirements that need to be fulfilled, in order to completely define this modelling method. We thus come to a formal description that covers these.

3.2.1   Formal Description

Unfortunately, it is not easy to fix a unique formal definition for the term "agent-based modelling", in order to be able to make a clean decision if a given theoretical model should an agent-based or not. This is due to the "problem" that agent-based modelling (as well as many other modelling approaches) is a scientific method that used in a multitude of disciplines; as a correct formal definition of a method is always linked to the corresponding scientific termini of a field of application, problems occur if the same method is used within disparate academical areas. To illustrate this point further, a definition of a method developed by mathematicians, given in a mathematical language, would not satisfy the needs of computer scientists or social scientists (and vice versa). In the specific case of agent-based modelling, scientists from different fields have developed their own formal descriptions, using their own scientific termini. It is obvious that this has cased a lot of communication problems, especially at interdisciplinary conferences. In order to solve these problems, very rough definitions were developed, sometimes even directly within the context of those conferences by interdisciplinary discourse. One of these definitions, developed at the Winter Simulation Conference 2005 and 2006 (Macal and North 2006), states that:

Definition 3.1.   An agent has to

- be uniquely identifiable
- cohabitate an environment with other agents, and has to be able to communicate with them
- be able to act targeted

- be autonomous and independent
- be able to change its behaviour

The principle meaning of these five points is now explained in full detail:

- *Unique Identification*: Although different agents in an agent-based model do not necessarily have to have different properties, they have to remain distinguishable at any time during the simulation. More specifically, all actions of one agent have to be traceable after a simulation run.
- *Environment*: All agents of a model share an environment in which they can interact both with each other and "the world" around them. The most common way of doing the latter is movement. In fact, most computer scientists define the property "movement" of an agent as a basic requirement for agent-based modelling. However, the general definition developed at Winter Simulation Conference (2005 & 2006) states that there may also be non-movable agents living in a totally abstract environment. As example for this, one could imagine geographically fixed subsidiaries of a company interacting on an abstract electronic network (see Example 3.3).
- *Act Targeted*: Within the environment, agents have to act so as to reach a certain goal or purpose. So, one could say that they have to be given a basic form of "intelligence".
- *Autonomous and Independent*: It is a very difficult philosophical question how the independent and autonomous action of an agent is justified, if any action/reaction rule is predefined by the modeller himself. The point of the definition can be interpreted in the following way: If the fictive agent would be given a personality without knowing its creator, the agent has to think of its own actions to be performed *independently from the other agents*.
- *Behaviour*: Evoked by its target (see previous point "Act Targeted"), every agent displays its own behaviour regarding communication and actions. If necessary the agent has to be allowed to change it.

Although these five ideas might pose a sufficient definition to modelling experts already knowing about agent-based modelling, non-experts will probably be no wiser now than they were before. Regarding modelling and simulation, theoretical and abstract definitions usually do not provide any concept on how to practically convert these theoretical ideas into useful and working models, as they are solely developed to determine the affiliation or non-affiliation of a given model to a modelling technique. Especially for scientific papers and conferences, this kind of classification is essential. In order to satisfy the expectations of the (probably disappointed) reader, the modelling approach is now presented in a little bit more user-friendly way, by using a step-by-step instruction. This is done via directly developing some example models in the next sections.

Fig. 3.3 Agent-based approach towards modelling a road junction, as given in Example 3.1

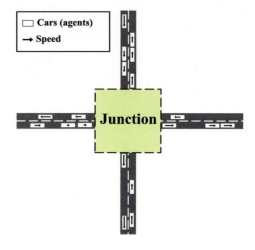

Example 3.1. Road Junction

Problem: A new road junction bringing together four streets (also see Fig. 3.3) is to be planned. In that context, the responsible traffic planners have to decide on either building a classical junction with traffic lights or to employ a roundabout. The stated goal is to avoid traffic congestions as far as possible. Experiments are not possible, so a modelling expert is called in to answer on this question.

Abstraction: Analysis of the situation reveals that a microscopic modelling approach is preferred, as the system can be reduced to consist of cars (active parts of the world) and streets (passive). Since colour, shape and type of the cars can be neglected, they can be reduced to appear as line-shaped (one-dimensional) objects with a certain length, moving individually on streets that are also line-shaped.

Model: Each car is represented by exactly one agent having a certain length. To enable interaction between the agents, an environment has to be defined: Each agent is assigned a place on one of the four one-dimensional streets (to be precise, each agent occupies a spatial interval of its own length). Additionally, each agent is assigned an initial velocity and a spatial target (i.e. one of the four ends of the streets, which he tries to reach as fast as possible). To control interactions between the agents, one has to furthermore define how each agent adjusts its velocity to avoid collisions with other agents while at the same time obeying the rules of the traffic lights or roundabout. In our case, both of these two mentioned points happen passively, by letting a would-be obstacle or traffic light send out: "avoid colliding with me".

(continued)

Example 3.1 (continued)

Conclusion: The developed agent-based model suits all five points mentioned earlier. Individually behaving agents (cars) are interacting in a goal-directed manner, within a specified environment—four linked one-dimensional lines (streets). Eventually, the success of the model will depend on the quality of the defined rules and the parameterisation. As a side-note, what we have just given as example is inspired by the famous Nagel-Schreckenberg model for simulating traffic and congestions on a one way street (Nagel and Schreckenberg 1992).

It can be observed that the strategic success of the modelling process before was precipitated by consequently answering the following questions:

1. Which parts of the system are expressed by agents?
2. What are the states of the agents? How do agents "look like"?
3. How does the environment "look like"?
4. How does communication between the agents take place? How and when do agents change their state?

To demonstrate the efficiency of these impulse questions another example is given in Example 3.2.

Example 3.2. Cave Lion

Problem: The extinction of the European cave lion (*Panthera leo spelaea*, see Christiansen 2008) is investigated. Archaeozoological findings prove the existence of the species in Europe until the onset of the first Lateglacial Interstadial, which occurred around 14.7 ka cal BP (see Stuart and Lister 2011). A possible hypothesis for the extinction is the "Keystone Herbivore Scenario" (Owen-Smith 1989), which states that the predator became extinct because of the extinction of its prey species. A second hypothesis goes into the direction of a reduction in prey numbers, possibly also connected to a geographical dispersal (Stuart and Lister 2011, p. 2337). In order to investigate these two hypotheses, a modelling expert is asked for help.

Abstraction: Based on estimations of the amount of lions and their prey, the key to an answer lies within the coexistence of two life forms, the lions and a prey species (i.e. reindeer and young cave bears, according to Bocherens et al. 2011; however, for the sake of the simplified example given here, we

(continued)

Fig. 3.4 Agent-based approach (see Example 3.2) modelling two cohabitating species, cave lions (predators) and reindeer (their prey)

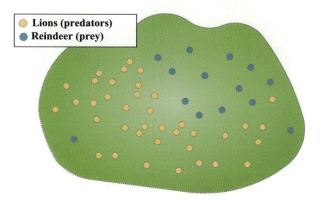

Example 3.2 (continued)

assume that the diet consisted only of reindeer). Most of the other environmental factors can be reduced to influences on the birth and death rate (directly or indirectly). As macroscopic approaches are usually unhandy regarding the extinction of single animals, and consequently, a microscopic approach is chosen.

Model: Each reindeer and each lion is represented by one agent. All reindeer agents avoid predators as long as possible and try to reproduce themselves. Each lion tries to catch prey in order to survive and to reproduce as well. The two classes of agents can be simplified to being dots on a map (see Fig. 3.4) trying to chase themselves. If height difference does not influence the system, a two-dimensional map of the area is a sufficient representation of the environment. Finally, rules concerning movement, reproduction/death and the so-called "chasing radius" for all lions complete the model definition.

Conclusion: The presented model with two different classes of agents defined here representative of a class of models called Predator-Prey models. This specific model was inspired by the "Wa-Tor" model (Dewdney 1984) simulating sharks and fish on a toroidal "planet". There also exist macroscopic models for the same purpose, for example Lotka-Volterra model based on first-order non-linear differential-equations (Lotka 1956).

To finally show that the environment can also be of very abstract manner, a third and last example is shown.

Fig. 3.5 Agent-based
approach modelling a number
of archaeological sites, as
given in Example 3.3

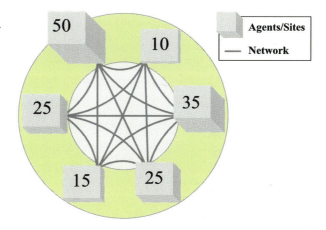

Example 3.3. Network of Archaeological Sites

Problem: A large team of archaeologists is distributed to six prospection sites.
The distribution of archaeologists should be solved optimally: Each site
has a certain capacity, which may not be surpassed. On the other hand,
each site demands a certain amount of people working there, or else the
prospection cannot be conducted effectively.

Abstraction: A first observation shows a positive correlation between the
number of archaeologists of a site and the corresponding work output (this
is, until the site gets too crowded for everyone to work efficiently, i.e. the
capacity is surpassed). The correlation coefficients and the capacities differ
from site to site. For the sake of this example, we furthermore assume
that these factors are independent from individual qualifications of each
archaeologist. Although the system itself is not dynamic (everyone keeps
working at his/her originally assigned site), it can be that the results of
a dynamic simulation in which archeologists are relocated might help to
determine the optimal distribution of workers.

Model: The "obvious" option to represent each archaeologist and site as agent
would work well in situation. However, this time, we choose to model
only the sites as agents (see Fig. 3.5) and the archaeologists as *the state
of these agents*, in the following manner: Each site stores the number of
archaeologists being occupied there, the demand and the capacity (both in
terms of people). Since sites are abstract, a detailed specification of the
environment is not necessary. All sites should be in contact at all times
during the simulation (e.g. by mobile phone, email or any other form of
communication), which leads to the introduction of links between each site

(continued)

Example 3.3 (continued)

   (see network of edges in Fig. 3.5). Using this communication infrastructure, the sites can arrange an exchange in personnel so that the workload is handled dynamically, using a "supply and demand" approach.

Conclusion: If done properly, finally an equilibrium state should be achieved that distributes the research team. This model is inspired by economic models presented by Aoki (2002), which shows once more that agent-based modelling is more than dots moving (and colliding) in a spatial environment.

As demonstrated, the field of application for agent-based models is very broad. Especially if there is no scientific basis allowing white box models, agent-based approaches are getting more and more popular: They are visual, descriptive, easy to understand and very flexible. This is leading to a multitude of models not only differing in purpose and parameterisation, but also in model structure. Thus, it is sometimes necessary to subdivide the class of agent based approaches in support of a scientific metadiscourse about them. Usually, this is done by specifying a certain part of the model a little bit closer. The following sections therefore give a run-down of the most frequently used classifications in that sense.

3.2.2 Specification of Time: Discrete or Continuous

Agent-based simulations are usually dynamic simulations investigating and simulating the temporal behaviour of a system. This can be done either continuously or discretely (a fact that does not only apply to agent-based models alone, as every dynamic model can be assigned to one of these two classes). In order to be perfectly clear about these two termini "time-continuous" and "time-discrete", we now explain them for (ordered) sets. For those not interested in this rather mathematical excursus, it is recommended to skip to the definition time-continuous/discrete agent-based models further down in the text.

It may sound trivial, but a *set* is no more than a simple assembly of any number and kind of items, called *elements* of the set. Usually the usage of curly brackets (set braces) indicates the assembly of all items in between. The order, how the elements are assembled is not important. Examples for sets:

- Room = {Couch, Table, TV, Chair}
- $A = \{1, 9, 4, 3, 2\}$
- Alphabet = $\{a, z, b, y, e, f, \ldots, m, p\}$
- NaturalNumbers = $\{1, 2, 3, \ldots\}$

Furthermore, a set is called *ordered* (or, to be precise, *totally ordered*) if every two different elements of the set can be compared and it can be determined which of them the bigger element is. For example, the previously defined set "Room" is (basically) not ordered, as it cannot be determined whether "Couch" is bigger or smaller than "TV". Obviously the sets "A" and "NaturalNumbers" are ordered. Although the set "Alphabet" is basically not ordered, instinctively one might say $a < b < c < \ldots$. So it is possible to define an order in a very natural way, making the set "Alphabet" structurally equivalent to the set of all Numbers from 1 to 26. As every (totally) ordered set is structurally equivalent to a set containing numerical values, only those sets are analysed further.

The word "continuous" derives from the Latin adjective *continuous* and means *to do something steadily without interruptions*. We now try to derive the idea of continuous sets in a very intuitive way, by comparing the following two totally ordered sets:

- $X = \{1, 2, 3\}$
- $Y = \{1, 1.2, 1.4, 1.6, 1.8, 2, 2.2, 2.4, 2.6, 2.8, 2.9, 3\}$

Set Y can sloppily be seen to be a little bit "more continuous" version of set X, as the gap (respectively the interruption) between two consecutive elements is smaller. More specifically, for a perfectly continuous set it is necessary that the gap between every two consecutive numbers has to become infinitely small. As a consequence, a continuous set has to contain an infinite number of elements. More precisely, it has to contain *every* real number in a chosen interval, which is written using square brackets:

- $Z = [1, 3]$

The interval Z is a continuous set containing 1, 3 and all real numbers in between. It also contains the sets X and Y defined earlier. Finally, a proper definition of continuity can be presented:

Definition 3.2. An (ordered) set is called *continuous* if for any two different elements of the set there is at least a third element in the set that lies in between them.

The set/interval Z fulfils this condition. Regarding continuous sets, it is important to note that it is not possible to determine a "next greater" element. In case of the previously defined set A, the next larger element of 4 would be 9. In case of the continuous set Z, it is not possible to determine which element of the set is e.g. the next-larger number after 1.5. It can not be determined *by definition*: In the hypothetical case where one has found the next bigger element after 1.5, the definition of the continuous set states that there *has to be* an element lying in between 1.5 and this number! Motivated by this observation, a definition of a discrete set is given:

Definition 3.3. An (ordered) set is called *discrete* if every element has exactly one next-larger element.

This leads to three important observations:

- All finite sets (i.e. all sets with a finite number of elements) are discrete.
- If a discrete set (containing positive numbers) is not finite the elements are not bounded and become bigger and bigger.
- If otherwise all elements of a discrete set are bounded, the set is finite.

Finally, a correct definition of a time-continuous and time-discrete simulation can be given:

Definition 3.4. If the set of time-steps used for a dynamic model is a continuous, the model is called *time-continuous*. If otherwise the set of time-steps used for the model is a discrete set, the simulation is called *time-discrete*.

Some further important facts:

- Neglecting quantum mechanics, time can be seen to proceed continuously in reality. Thus, time-continuous models are usually preferred to discrete models.
- Unluckily, it is impossible to perform time-continuous simulations. Thus, a model that was designed as being time-continuous designed has to be "discretised", i.e. simulated time-discrete. As a consequence, we need to differentiate between time-continuous/discrete modelling and time-continuous/discrete simulation.
- Agent-based models are usually time-discrete models with equidistant time-steps (e.g. seconds). The decision-making process is not continuous, but happens as series of *events*: At a certain time step, the agent "decides" to take an action, resulting in a sudden state change. Some of these discrete state changes are harder to spot—for example when looking at movement: Even though it may seem that an agent moves continuously, it really performs a series of "hops" (once per time step) that we wrongly observe as being continuous. Truly time-continuous agent-based models are very rare. From the three examples presented earlier, solely the car model (see Example 3.3) would be suitable.
- The time-set does not necessarily have to be *either* continuous *or* discrete: There are also *hybrid models* where a mixture of partially continuous and discrete time intervals is possible.

3.2.3 Specification of the Environment: Discrete or Continuous

Definition 3.1 does not give any restriction regarding the shape and appearance of the environment, as long as it is inhabited by all agents. To illustrate this point further, the archaeological site model that was given as Example 3.3 did not include any topological aspects at all. Thus, we must first consider whether the environment contains topological concepts or not. If yes, a further classification into either spatially discrete or continuous can be conducted.

Fig. 3.6 Sketch of a spatial discrete compared to a spatial continuous agent-based model

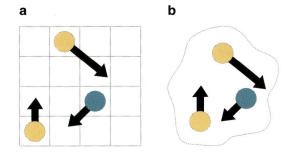

Figure 3.6 shows examples of a discrete and a continuous environment inhabited by agents. Spatially discrete models subdivide the environment, usually into equivalent cells as indicated in Fig. 3.6a. Such models are a lot easier to deal with, for instance because the calculation of movement happens in a finitely limited space.

Illustrating this point further using the predator-prey model previously presented in Example 3.2, it is clear that a reindeer chased by two lions should follow some strategy concerning an optimal direction of escape. Dealing with a spatial-discrete model, the optimal strategy can simply be determined to be one direction among a finite number of possible directions, depending on the position of the two predators. In a spatially continuous model, an infinite number of directions has to be taken into account. Likewise, distances and neighborhoods are also more complex to compute: Spatially discrete models can count the number of cells between agents and get all agents "around" a given agent (either the Von Neumann neighborhood consisting of the four cells north, south, east and west, or the Moore neighborhood additionally consisting of the north-east, north-west, south-east and south-west cell). Spatially continuous models must instead evaluate the distances (what is far, what is near?) and introduce radii (which radius is best?) around an agent.

Example 3.4. Salt Crystallisation

Problem: In order to investigate the crystallisation process of salt in a mine, a liquid model which is to be designed. Common liquid models use rather difficult mathematical formulae (e.g. Navier-Stokes-partial differential equations), which do not seem to work within this special case of modelling fine crystalline structures. Therefore, a microscopic approach is used. As we will see, this is a mixture between an agent-based approach and Cellular Automata (CAs), another type of microscopic approach[2] (also refer to Fig. 3.7).

(continued)

---

[2]The transition between agent-based modelling and CAs is rather fluent: One could say with some caution that "agents" are potentially movable entities, while Cellular Automata specify a discrete environment of grid cells having rules that change their state (see e.g. Conway's "Game of Life" reported in Gardner 1971 and Berlekamp 2001).

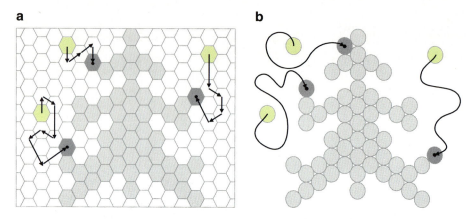

Fig. 3.7 Spatial discrete and continuous salt crystal model (see Example 3.4)

Example 3.4 (continued)

Abstraction: Observations done by microscope show on the one hand a rather random movement of dissolved ions in the liquid (Brownian motion) and on the other hand a very systematic crystallisation process happening at certain angles around the crystallisation seeds. The corresponding ions are thus abstracted as being agents behaving according to a Brownian motion in the liquid, finally sticking to one of the crystallisation seeds under certain conditions and at certain angles.

Model: The main task of this model is the specification of the environment and especially the crystallisation seeds and borders. This time, we model two complementary options: The systematic crystallisation process in which certain crystallisation angles are preferred can be implemented by using a spatially discrete model utilizing a hexagonal grid (see Fig. 3.7a). Whether one would like to call that a CA (a cell has its own behaviour for propagating crystalisation to the next cell) or an agent-based model (agents moving from cell to cell in a discretized environment) is a philosophical question leading to the same result. On the contrary, only a space-continuous model allows a correct implementation of a random-walk process (Brownian motion), which can be seen in Fig. 3.7b.

Conclusion: Typically, crystal growth is nowadays simulated by CAs. However, as this example shows, one may also choose a continuous approach, and both model types complement themselves nicely. Another point for consideration is that there is not always "a best approach", but possible a multitude of angles from which a problem can be approached. The context of this model, crystallisation, shows how simple microscopic rules can lead to a very complex, aggregated behaviour. For more detailed information about this model, see Packard (1986).

3.2.4 Stochastic and Deterministic Agent-Based Models

To give a final classification, the difference between stochastic and deterministic agent-based models is given. The word "stochastic", translated from original Greek, means "the art of probability" or "the art of randomisation". Thus, a model is generally called stochastic if the model is designed to contain random parameters or variables.[3] Such kinds of models have to be used if there is a lack of detailed information or uncertain data about specific modelling parameters. Another case is when a process cannot be modelled to such a degree of detail so as to be simulatable deterministically (also see Example 3.5).

When a model's inner mechanics are known (white box approach), it is easy to find out it contains randomisation, thus making it stochastic. But what if we have a black-box model? In this case, we can run a model twice with the same input/parameters and observe its output: If the outputs are the same, the model is deterministic, in all other cases stochastic. More formally:

Definition 3.5. A model is called deterministic, if and only if one initial configuration and parameterisation always leads to one specific simulation result. Otherwise the model is called stochastic.

Some further facts:

- Stochastic models must be simulated numerous times, until the obtained results are significant (e.g. within a confidence interval, variances, etc.). If this is not possible, simulation must be performed as often as possible (Monte Carlo Method). For agent-based models, this usually translates to high computational requirements.
- In all cases, obtained results demand proper interpretation by experts.
- Surprisingly, deterministic models sometimes lead to far more complex behaviour than stochastic ones. A perfect example is Conway's Game of Life (reported e.g. in Gardner 1971; Berlekamp 2001), in which very simple deterministic rules lead to a nearly chaotic behaviour.

Example 3.5. Pool Billiard

Problem: Pool billiard is a physical game: Success seems to not dependent on luck, but rather on controlling the angle, spin and power of a ball.
Abstraction: The same angle, spin and power should lead to identical shots. Therefore, a deterministic model would seem to be an adequate solution in this case. On the other hand, pool-experienced reader might insist that

(continued)

---

[3]As a side-note, stochasticity is not a special property of agent-based models, but applies to all kinds of models.

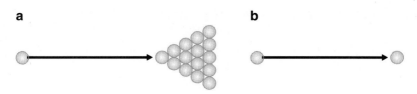

Fig. 3.8 It is not possible to say whether deterministic models or stochastic models are the best approximation for reality (refer to Example 3.5)

Example 3.5 (continued)
  the first shot (also called break-off shot, see Fig. 3.8a) contains lots of "random" effects and is usually not predictable. This is due to the fact that lots of factors influencing the 15 colliding balls (e.g. dirt-particles, surface irregularities and even quantum-mechanic effects) can hardly be measured before the shot.
Conclusion: It is not possible to say whether a deterministic (Fig. 3.8b) or stochastic model are the best approximation of reality in this specific case: Multi-body collisions behave stochastically, single-body collisions deterministically.

3.3 General Advantages and Disadvantages of Agent-Based Models

Out of all benefits that agent-based modelling brings, the ability to describe swarm behaviour is probably the most extraordinary one. It is mathematically a very surprising observation that the aggregated behaviour of a cohabitating group of agents is extremely difficult to be predicted when only investigating the cohabitation rules and parameters of the agents: This swarm behaviour that emerges unpredictably from a great number of interacting individuals (found e.g. with ants, migrating birds or humans at a soccer stadium) is now covered in more detail.

3.3.1 Emergent Behaviour

The term *swarm intelligence* or *swarm behaviour* was first introduced by Beni and Wang (1989) within the context of their work in robotics. It means that the predictable and simple behaviour of individuals leads to an unpredictable, complex behaviour for the whole group (also refer to Fig. 3.9). Today, we also use the terms *crowd intelligence* or *crowd behaviour* synonymously.

Although emergent phenomena arising from a large number individuals are well-known since long time, research in this field is still young (e.g. pedestrian and

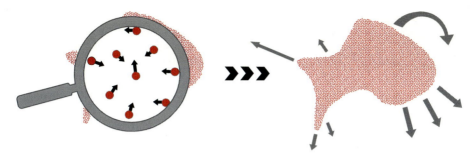

Fig. 3.9 Basic idea of emergent crowd behaviour respectively swarm-intelligence. The simple behaviour of individuals leads to (*shown left*) an unpredictable, complex behaviour of a whole crowd (*shown right*)

$$\frac{\partial \Theta(\vec{x}, t)}{\partial t} = A(\vec{x}, t) \triangle_{\vec{x}} \Phi(\vec{x}, t)$$

Fig. 3.10 "Intuitive" agent-based versus theoretical description of a system

evacuation simulation). On the one hand, it might be that the complexities of this topic have led to a slow take-up by the academic community. On the other hand, everyday computing has only recently reached performance levels at which large scale simulations are feasible. In that context, agent-based simulation is not simply used for "simulating behaviour", but as investigation tool for crowd intelligence.

3.3.2 Natural Description of a System

An agent does not know about the behaviour of the whole group but instead focuses on itself and its surroundings. Thus, the necessary level of abstraction needed is very low. This aspect has a lot of benefits: It enables easier interdisciplinary discourse between the creators of the models (technical disciplines and professionals such as archaeology) and their audience, with an emphasis on "who does what". This storytelling approach (as we would call it) may overcome the huge differences in mathematical skills: At least the structure of agent-based models can be explained very easily, the details are hidden in the code underneath (also see Fig. 3.10[4]).

---

[4]The diffusion equation at the right hand side of the picture only implies a model developed with difficult mathematical formulas. The diffusion equation is usually not used to model cohabitation of ants.

The intuitive representation of systems within microscopic approaches as well as the freedom granted by that modelling technique lead to extremely flexible models. Re-parameterisation as well as structural changes can be accommodated easily, by changing only a few lines in the implementation. From a computational design standpoint, this is called "Separation of Concerns" (SoC): Every part of the system behaves in a modular fashion, exposing information relevant to the outside (in our case e.g. the observable movement) but hides all further inner mechanics. Of course, this also implies that bugs hidden deep inside the code are hard to be found—which is one of the disadvantages mentioned in the next section.

3.3.3   Dangers and Problems

There are also some disadvantages in agent-based modelling, which can not all be attributed to the short history of the modelling method but may extend further to its underlying concept. On the surface, two important observations can be made when critically analysing agent-based models:

1. All agent-based simulations are computationally expensive. Doing serious simulation work, we may need millions or even billions of agents in order to receive useful results. So, these simulations might require a lot more powerful technologies than just personal computers—which is when high-performance computating (also see Chap. 6) comes into play. For example, parallel-computing (spread calculations onto several kernels), utilisation of the Graphics Processing Unit (GPU), cloud- and server-computing might all be used to make a simulation feasible, the decision over what method to take is, however, different in every model.

2. As mentioned before, the flexibility of the modelling technique also involves dangers: As every model is designed for a certain purpose, its field of application and especially validity is limited. Too many modifications within the model structure and wrong interpretation of the output can lead to wrong and invalid modelling result. One may also be tempted to interpret a model beyond its *border of validity*. Taking the Cave Lion example as a reference (also see Example 3.2), we may recall that this was a predator-prey model designed to predict the temporal development of the two cohabitating species. If the model is validated and calibrated the right way, it is correct to predict a certain number of animals at a specific time also in the real system. However, it is *not correct* to assume that the emerging hunting strategies of the model map to the hunting strategies in the real system, as they are (usually) too simplified!

Clearly, the "individual-based" simulation approach is what lies at the core of both scalability problems as well as dangers when every part comes together (overseen problems in the individual ruleset leading to a false aggregated result—a situation which is complex to debug). Even if modifications and extensions within the model are done properly, they usually introduce a lot of new modelling assumptions and parameters, which have to be calibrated and tested. Concluding, there are hardly any modelling techniques as flexible as agent-based modelling, even though they are reacting quite sensitive to modifications and oversights in the specification of individual behaviour. Validation of these kind of models is a real challenge and requires a lot of time and effort. Thus, an own chapter is devoted to exactly this topic (see Chap. 4).

References

Aoki M (2002) Modeling Aggregate Behavior and Fluctuations in Economics: Stochastic Views of Interacting Agents. Cambridge University Press, New York

Aristoteles (1985) Metaphysics: Zeta, Eta, Theta, Iota. Books VII-X. HPC Classics Series. Hackett, Indianapolis. Edited by M. Furth

Beni G, Wang J (1989) Swarm Intelligence in Cellular Robotic Systems. In: Proceed. NATO Advanced Workshop on Robots and Biological Systems, Tuscany, Italy

Berlekamp ER (2001) Winning Ways for Your Mathematical Plays, 2nd edn. A.K. Peters, Natick, MA

Bocherens H, Drucker DG, Bonjean D, Bridault A, Conard NJ, Cupillard C, Germonpré M, Höneisen M, Münzel SC, Napierala H, Patou-Mathis M, Stephan E, Uerpmann HP, Ziegler R (2011) Isotopic evidence for dietary ecology of cave lion (Panthera spelaea) in North-Western Europe: Prey choice, competition and implications for extinction. Quaternary Int 245:249–261

Christiansen P (2008) Phylogeny of the great cats (Felidae: Pantherinae), and the influence of fossil taxa and missing characters. Cladistics 24(6):977–992. DOI 10.1111/j.1096-0031.2008.00226.x

Dewdney AK (1984) Computer recreations: Sharks and fish wage an ecological war on the toroidal planet wa-tor. Sci Am, 14–22

Gardner M (1971) Mathematical games: On cellular automata, self-reproduction, the Garden of Eden, and the game "life". Sci Am, 112–117

Helbing D, Molnár P (1995) Social force model for pedestrian dynamics. Phys Rev E 51(5): 4282–4286

Helbing D, Farkas I, Vicsek T (2000) Simulating dynamical features of escape panic. Nature 407:487–490

Kant I (1784) Beantwortung der Frage: Was ist Aufklärung? In: Gedike F, Biester JE (eds) Berlinische Monatsschrift. Haude und Spener, Berlin, pp 481–494

Kant I (1787) Werkausgabe, vol 3, 4th edn, Insel Verlag, Frankfurt, chap Vorrede zur zweiten Auflage, p 23. Edited by W. Weischedel

Lotka AJ (1956) Elements of Mathematical Biology. Dover, New York

Macal CM, North MJ (2006) Tutorial on Agent-Based Modeling and Simulation Part 2: How to Model with Agents. In: Proceedings of the 2006 Winter Simulation Conference, Monterey, California, pp 73–83, URL http://www.informs-sim.org/wsc06papers/008.pdf

Moore GE (1965) Cramming more components onto integrated circuits. Electronics 38(8):114–117

Nagel K, Schreckenberg M (1992) A cellular automaton model for freeway traffic. J de Physique I 2(12):2221–2229. DOI 10.1051/jp1:1992277, URL http://www.edpsciences.org/10.1051/jp1: 1992277

Owen-Smith N (1989) Megafaunal extinctions: The conservation message from 11,000 years b.p. Conservat Biol 3(4):405–412

Packard NH (1986) Lattice Models for Soldification and Aggregation. In: Proceedings of the First International Syposium for Sience on Form. KTK Scientific Publishers, Tokyo, pp 95–101

Predtetschenski WM, Milinski AI (1971) Personenströme in Gebäuden - Berechnungsmethoden für die Projektierung. Verlagsgesellschaft Rudolf Müller, Köln-Braunsfeld

Rogsch C, Klingsch W (2011) To See Behind the Curtain - A Methodical Approach to Identify Cal-
    culation Methods of Closed-Source Evacuation Software Tools. In: Proceedings of Pedestrian
    and Evacuation Dynamics 2011, pp 567–576
Rogsch C, Weigel H, Klingsch W (2008) Hand-Calculation Methods for Evacuation Calculation -
    Last Chance for an Old-Fashioned Approach or a Real Alternative to Microscopic Simulation
    Tools? In: Klingsch W, Rogsch C, Schadschneider A, Schreckenberg M (eds) Pedestrian and
    Evacuation Dynamics 2008, pp 523–528
Shannon RE (1975) Systems Simulation: The Art and Science. Prentice-Hall, Englewood Cliffs,
    NJ
Stachowiak H (1973) Allgemeine Modelltheorie. Springer, Wien, New York
Stuart AJ, Lister AM (2011) Extinction chronology of the cave lion Panthera spelaea. Quaternary
    Sci Rev 30:2329–2340. DOI doi:10.1016/j.quascirev.2010.04.023
Wolfram S (2002) A New Kind of Science. Wolfram Media, Champaign, IL

Chapter 4
Reproducibility

Niki Popper and Philipp Pichler

4.1 Introduction

One of the fundamentals of scientific work is that knowledge should be *transparent*, i.e. openly available for professional discourse. In this chapter, we specially deal with this aspect from the angle of *reproducibility,* focusing on how to give and gain input from fellow researchers. From the side of the project team, this demands a statement of limitations and assumptions within a model. As a matter of fact, possible shortcomings will be detected and assumptions may be questioned. This is no reason to be intimidated, however, as this process pays off in the long term.

Reproducibility is a challenging task and can be cost intensive. Thus, all efforts that help to achieve it should be carried out with respect to their benefit. Special attention should be paid to documentation, visualisation, parameter formulation, data preparation, verification and validation, most of which will be presented during the next pages. One exception is a detailed presentation of data preparation, which would constitute a chapter of its own. We may instead forward the interested reader to Freire et al. (2012). Another topic that is not covered in full detail is how a model structure can be communicated. Recommendations on how to do this correctly and efficiently were, for example, given by the ODD protocol (Grimm et al. 2006), which we give as an overview (see Sect. 4.5.1). The legal aspect of communicating model structure, as covered e.g. in Stodden (2010), is another topic which is left out intentionally.

N. Popper (✉) • P. Pichler
DWH Simulation Services, Vienna, Austria
e-mail: niki.popper@dwh.at; philipp.pichler@dwh.at

© Springer International Publishing Switzerland 2015   77
G. Wurzer et al. (eds.), *Agent-based Modeling and Simulation in Archaeology*,
Advances in Geographic Information Science, DOI 10.1007/978-3-319-00008-4_4

4.2 Outline

Readers will gain a better feeling for what is important to achieve reproducibility in the following sections: We start with a general description of the development process behind a modelling and simulation (M&S) project. Understanding this *lifecycle* is a pre-condition for talking about reproducibility, since one needs to know exactly in which phase what information is produced.

The next sections deal with parameter formulation, documentation and verification/validation. These topics help to substantiate that the developed simulation produces reliable and useable results. These can then be used to gain knowledge that confirms hypotheses—rectification—or to identify wrong hypotheses—falsification. In archaeology, a huge collection of hypotheses is possible, due to missing information. Falsification of assumptions and hypotheses reduces the amount of possibilities. We will thus take a closer look at this aspect when concluding.

4.3 Lifecycle of a Modelling and Simulation Study

The process of developing a model and implementing it as a simulation can be referred to as its *lifecycle*. To understand reproducibility requires a careful look at this subject, because we first need to define its basic constituents—phases, contained concepts and resulting deliverables, that are later being referred to. In general, an M&S project does not evolve in a straight-forward manner, but rather iteratively (in a *spiral process*): The model/simulation is redefined several times, until it can be determined to work correctly for its preset goal. In that context, it is noteworthy to say that a model is concerned only with a (simplified, limited) portion of reality. Modellers have to make abstractions, assumptions and define boundaries to get to an easier view—though detailed and complete enough for the study question— that is computable. That is why the whole process might have to be done several times, until the right assumptions and abstractions are made. We now give a brief description of the basic structure of the M&S lifecycle, before coming to a more narrative description of the same matter.

Figure 4.1 presents a generalized view of a M&S lifecycle on the basis of work done by Sargent (2010), with some slight adaptions[1]:

- A *problem* arises and is being formulated as one or more study questions, which guide the development into the right direction. The ultimate aim of the M&S project is to solve the stated problem and answer on these defined study questions.

---

[1]"Knowledge Building and Interpretation" was added.

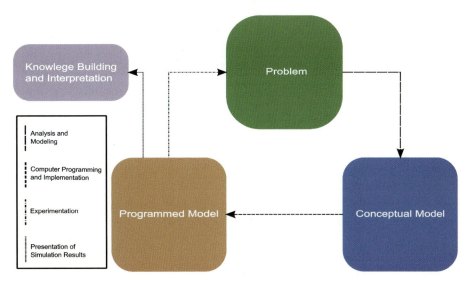

Fig. 4.1 Adaptation of the "Generalized Lifecycle of a Simulation Study" given by Sargent (2010)

- In a next step, the system is analysed and modelled, which leads to a *conceptual model* that can solve the problem and give answers on the study questions.
- The *conceptual model* is then implemented in some programming language, leading to a *computerized model* which can either produce new findings (leading to a redefinition of the problem and thus resulting in a new iteration) or produces results that are validated and verified (see Sect. 4.6), thus being credible.
- Using such a credible model, developers, experts and users may produce results that reflect reality *within its predictive boundaries* and calculate possible scenarii, which can be used in the context of *decision support*.

Figure 4.2 shows a more detailed version of the same lifecycle, based upon Balci (1994):

- The problem phase is split into the communicated problem, formulated problem, proposed solution technique and system and objectives definition.
- The conceptual model phase is divided into the conceptual model and the communicative model.
- The computerized model phase furthermore includes the programmed model, the experimental model and the simulation results.

We now give a brief example of this project structure, using the case of a hypothetical project as a basis. Note that the description is intentionally held brief and would be much larger (an understatement: it could rather fill a whole chapter) for a real project.

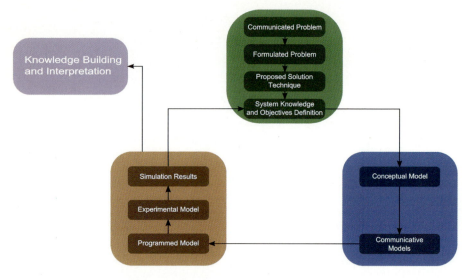

Fig. 4.2 Detailed Lifecycle of a Simulation Study, based on Balci (1994)

Example 4.1. Lifecycle for an Army Model

Communicated Problem: In the context of campaign analysis, archaeologists want to look into the process of moving roman foot-soldiers between Messana and Syracuse in the first Punic War (264–241 BC).

Formulated Problem: We formulate the problem as a set of study questions, e.g.: How fast were the foot soldiers marching (time span, speed)? How far in one day?

Proposed Solution Technique: Upon examining the problem closer, we find that it is indeed applicable for an M&S study. Moreover, we opt for agent-based modelling, since the influence of the terrain and the interaction between individuals is probably influential for the overall model.

System Knowledge and Objectives: Foot soldiers marched as a *legion* of approximately 4,800 soldiers. The line-of-flight distance between Messana and Syracuse is 100 km (assumption). By foot, this distance depends on the terrain (which is taken from Geographical Information System).

Conceptual Model: With the given prior knowledge, the modeller thinks of an agent-based model in which a legion is represented as one agent. The terrain is represented as a discrete grid, with a large cell size (50×50 m).

Communicative Models: We prepare two different models in written form: (1.) The mathematical view on the problem, intended for later implementation and (2.) the archaeological view, intended for discussion. The first contains

(continued)

Example 4.1 (continued)

a description of movement rules, speeds and interaction specifications between legions (there might be more than one), as formulae and definitions. The second communicative model gives a run-down of archaeological descriptions and mental models, which were used as basis for the formulation of the mathematical model.

Programmed Model: Based on the mathematical communicative model, a programmer implements a simulation in Java.

Experimental Model: Together with archaeologists, the modellers define several scenarii (i.e. sets of different parameter values[2]). The simulation is then executed multiple times with these.

Simulation Results: Executions of experimental models lead to results, which (in this case) are given as spreadsheet. These are then subject to interpretation (e.g. in comparison to historical data on this campaign), in order to find out if the results are sane.

After this step, the lifecycle may re-iterate (refinement). If refinement is not needed, then the results can be taken as current working model (see step "Knowledge Building and Interpretation" in Fig. 4.2).

Each part of the lifecycle has its own data and information requirements and leads to a certain generated output. What we really "reproduce" is the output at some stage of the lifecycle, and thus, a closer view is given in the next section.

4.4 Parameter and Output Definition

The basis of every M&S study lies in the information on which it is based (e.g. studies, databases, expert knowledge, statistical evaluations). In this context, we may differentiate between *data* and *general information*: Data contains all quantifiable information and is characterized as having a high degree of objectivity (e.g. as in values within a database). General information is the conglomerate of non-measurable information and has a high degree of subjectivity (e.g. expert knowledge). Regarding the lifecycle presented before, we may say that in each of its stages, a model/simulation transforms data and general information into an output, which is used by subsequent stages as input.

Data and general information enter the model in the form of *parameters*,[3] through a transformation process: When thinking of the whole M&S lifecycle,

---

[2]In reality, a scenario can also contain variations of algorithms, e.g. movement rules.

[3]The term "parameter" is used in different meanings across the disciplines. In this chapter, the mathematical/informatical view is presented.

each of the models contained in the different stages demands different parts of the data/information present. The subsequent "execution" of the model at a specific stage may not be a "simulation run" in the sense of a program execution. Rather, it may involve knowledge building and crystalisation (as for example in the Systems Knowledge and Objectives phase), resulting a derived output that is fed into the next stage.

Characterizing this even further (and with a view towards Agent-Based Simulation), we can say that each is carrying a name (e.g. "walking speed"), a data type (e.g. a number or Boolean), optional metadata (e.g. constrained to be in the range between 0-2 m/s) and a value (1.4 m/s). For the latter, we may further observe that:

- A parameter's value is *constant during the execution of a simulation*: For example, one might specify a walking speed for all agents that the simulation will use directly. But one might also specify this walking speed indirectly, giving minimal and maximal walking speeds (both are parameters in this case!) from which the simulation will derive the actual speed for each individual agent from a distribution (e.g. uniform distribution, normal distribution[4]).
- A parameter's value is *variable before the execution of a simulation*: It might be changed to experiment with different settings of the model, resulting in different outputs (always the same output for the same parameter settings: deterministic model; different outputs for the same parameter settings: stochastic model).

The set of all possible parameter settings is called *parameter space*. The act of choosing a value for a parameter is either called *parameterisation* and *calibration*: In the first case, we make a (hopefully well-informed) choice of a value, whereas in the second case, we pick the set of parameter values that have shown to be in good agreement with the reality that the model is concerned with. Put differently, parameterisation happens *a priori* and calibration *a posteriori*.

Figure 4.3 shows an extended version of the lifecycle presented earlier, with emphasis on definition and transformation of parameters, their parameterisation and calibration:

- The modelling lifecycle begins with the collection of general information and data. Both contribute to the System Knowledge and Objectives phase, in which they are transformed into structural knowledge about the system in question. Interestingly, the data values themselves are not important at this stage; the only thing that is important is to determine which information could be useful for the model (i.e. a first step towards defining its *boundaries*).
- For the conceptual model, the modeller thinks of a structural concept, separating structural knowledge into information that the model needs or produces. In other words, the modeller tries to establish the dependencies between the types of information that the model will likely use. A detailed definition of this information in the form of concrete parameters is, however, deferred to the

---

[4]Furthermore, a distribution might be discrete or continuous.

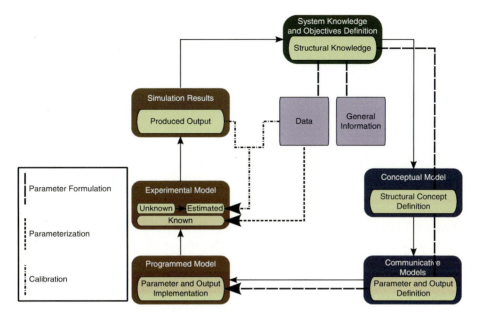

Fig. 4.3 Parameter formulation, parameterisation and calibration

communicative models (there might be several, as mentioned, each serving one specific audience): Guided by the gathered general knowledge and the format of the data, this phase produces a well-defined (minimal) set of parameters that can answer the problem. In that context, abstractions and generalisations are applied. For example, walking speeds might be assumed as uniform, if modelling soldiers marching in unison.

- Once proper communicative models with a minimal set of parameters and outputs have been developed, the implementation can start. The resulting programmed model transfers the (mathematical) notion of parameters and outputs into programming-language specific constructs (for example, the Boolean values given as "yes" and "no" are mapped to "true" and "false" in the source code).

As intermediate summary of these three points, let us again consider a walking model for a Roman army.

Example 4.2. Roman Army Model: From Data to Programmed Model

Let us suppose that a roman army had to walk from a city A to a city B. The question to be answered by the M&S study is: How far it did come after being on foot for a given amount of hours (i.e. a walking model).

(continued)

Example 4.2 (continued)

General Information and Data: The relevant literature is scrutinized for general information regarding the marching process in a Roman legion. From this source, we may derive that marching was highly coordinated and can be assumed to have been conducted in unison. Through experimental archaeology, walking speeds over similar terrain (e.g. regarding consistency and slope) are be established using people wearing reconstructed armour and weaponry. These are entered into a database (data!). Furthermore, the length of the (well-known) historical route between *A* to *B* is determined to be *n* km, based on calculations done inside a GIS package (also data!).

System Knowledge and Objectives: As established earlier, a legion walks in unison. Therefore, we can take the average speed of *b* km/h rather than individual speeds as basis for the movement. The objective can furthermore be more clearly formulated as: Given a certain walking time since departure at *A* (input parameter), how much is the remaining length until *B* (output).

Conceptual Model: If we accept that marching was done in unison, then it is possible to see the problem not as a discretized model of individually walking agents, but rather as continuous movement of a point along a line that connects city *A* to city *B*. As underlying method, we would then use the line equation to solve the model. Concerning parameters, there is also an obvious dependency between the spent walking time (superior) and remaining length (subordinate).

Communicative Models: Besides an archaeological model that sums up the relevant literature and assumptions, a more formal (mathematical) communicative model is written down. At its core, the model will be computed using the line equation $f(x) = n - b * x$, where $f(x)$ is the distance to city *B* in km (output), *n* is the length between *A* and *B* in km (parameter), *b* is the average speed in km/h (parameter) and *x* is the time in hours since departure from *A* (variable, i.e. parameter which is evaluated continuously to produce a simulation result). *n* and *b* are positive real numbers, *x* is in the range $[0, n/b]$ hours, $f(x)$ is in $[0, n]$.

Programmed Model: Using a programming language for actually coding the so-defined model, we can output a curve where the x-Axis is the time and y axis is the distance to B. Under the hood will need to make *x* sweep its whole defined range, using fixed intervals that we specify (e.g. $dt = 0.01$, which can be seen as additional parameter *given by the implementation*).

At this point in the lifecycle, the implementation has the capabilities to run a simulation. However, and in contrast to Example 4.2, we will likely not sweep through the whole range of parameter values possible. Rather, it is necessary to first find reasonable parameter values: The model is subjected to experimentation for exactly this sake, and becomes the "Experimental Model". Depending on the type of parameter under consideration, there are two different techniques for doing this:

- Parameterisation (known parameter values): Parameter values can be derived from (a prio known) data values. If the data cannot be transformed in a way that all parameter values can be inferred, further investigations are needed (i.e. broadening the scope of the data basis, gaining new data by conducting experiments, etc.). If this is not possible, some parameters have to be calibrated.
- Calibration (unknown parameter values): Values for unknown parameters are estimated, assisted by given data (statistical evaluations, studies or previous simulation runs). After the experimental model is run, the simulation output is compared to a possibly known output data or, as is often the case in archaeology, to constraints governing the output. If the simulated model produces an output that seems reasonable, the calibration task is finished and the parameter values are—in the best possible scenario—known. If on the contrary the simulation output does not fit, the unknown parameter values have to be estimated again and the experimentation process reiterated[5] (also refer to Example 4.3). If calibration fails, a redefinition of the model might be necessary and/or one of the known parameter values has to be doubted. The modelling process starts all over again, because the system knowledge might have changed.

Arguably, the act of calibration is positivistic and sometimes ill-suited for the archaeological field (little data to fit against!). Nevertheless, imposing constraints on what could be an output is in line with what the social sciences would consider a contribution in the first place; in that sense, we argue that the purpose of an M&S study for archaeology lies not primarily in the generation of results per se, but in an exclusion of unlikely scenarii that would otherwise enter the scientific thought process, leading to wrong conclusions. The definition of constraints is a positive "side-effect" of performing such a study, and may in fact lead to the realisation that a previously well-understood problem needs to be re-examined and clarified.

Example 4.3. Calibration

Estimated Parameter: The two cities A and B are estimated to be $n = 300$ km apart from each other.

Calibration: We have the information that the army was 260 km apart from city B after 10 h of walking. This information helps to calibrate the parameter b. Simulation runs are performed. If the value at 10 h is below 260 km, b is too low and has to be increased. If the value is higher, b has to be decreased. If the value at 10 h is exactly 260 km, the right parameter value has been found.

---

[5]This task is often supported by mathematical optimisation.

Similarly to the specification of parameters, outputs also need to be defined. For example, we may derive trajectories from a walking model, if the model is sufficiently detailed to be credible. We may also employ different representations for the same output. In the walking model, we may represent the trajectories in a more abstract fashion, using path-time diagrams, or very concretely, by superimposing paths on a map. This aspect rather belongs to visualisation, which is elaborated in due course.

4.5   Documentation

Reproducibility is strongly connected to documentation. While it might be easy to say that "all tasks that are important during the development of an M&S study should be written down", finding a way in which to represent an M&S study both accurately and efficiently is hard. There are at least three forms of documentation to choose from: Textual documentation, illustrations and, not surprisingly, the source code itself. We will now look into each of these three categories, giving an overview of techniques that might help in that context.

4.5.1   Textual Documentation

Probably every model author will have his own advice on how to produce textual documentation. Some examples that we found useful are:

Adequacy and Structure: The level of detail in the documentation should be proportional to the importance of the documented entity. Highly detailed topics should be structured into a hierarchy of gradually increasing complexity.

Simplicity: Whenever possible, documentation should be done in Basic English (Ogden 1940, 1968), a constructed language of 850 words which has a simplified grammar and is designed for the communication of complex thoughts. As a matter of fact, Basic English is easily translated (manually and automatically) and can serve a large international community as a basis for collaboration.

Clarity: A glossary or definition part may help to avoid ambiguities and shorten the text. However, this also carries the risk of over-formalisation. Examples can help to lighten up the text and illustrate an idea intuitively rather than formally.

Audience: Who needs to know what? There are several roles within an M&S project (e.g. users, model developers), each being interested in a different aspect (e.g. archaeological statements, technical descriptions). Making this explicit, possibly through use of formatting (italics, grayboxes, etc.) and labels, can help an interested user in picking out truly interesting information rather than being forced to read everything.

In addition to these general observations, there are also very specific guidelines for preparing documentation. The most widely known (quasi-)standard in archaeological simulation is the ODD (Overview, Design concepts, and Details) protocol by Grimm et al. (2006). We now come to a short description of its contained parts, according to the first revision given by Grimm et al. (2010), Polhill (2010):

Purpose gives a very short summary of the intent of the model.

Entities, State Variables and Scales names the types of agents or cells and their state variables. Furthermore, the spatial and temporal scale of the model is described.

Process Overview and Scheduling describes how the model is updated in each time step, preferably using pseudo-code. The authors also recommend to include a description of execution order (when multiple agents are involved).

Design Concepts consists of 11 sub-parts, some of which can be left away for small models or models in which they are not applicable: (1.) *Basic principles* outlines general concepts behind the model, (2.) *Emergence* gives a description of overall behaviour arising from the combination of individual behaviours, (3.) *Adaptation* states how individuals react to changes in themselves or their environment, (4.) Objectives gives a description of the goals followed, either by each agent individually or by a whole team of agents, (5.) *Learning* describes the change of adaptive traits as consequence of gained "experience", (6.) *Prediction* gives details over how agents judge a model's future state, (7.) *Sensing* gives an account of what internal and environmental state variables individuals can perceive and consider in their decision process, (8.) *Interaction* describes how agents affect each other and the environment, (9.) *Stochasticity* explicitly states if the modeled processes include a notion of randomness, (10.) *Collectives* gives details over how aggregates of agents are formed and (11.) *Observation* states what data are collected from the model.

Initialisation is concerned with the initial state of the model.

Input data lists required data from external sources, such as files, databases or other models.

Submodels lists all processes briefly outlined under Process Overview and Scheduling, in full detail, including parameterisation and an account of how such a submodel is tested.

According to Grimm et al. (2010), ODD can be overdone for very simple models. In this case, the authors propose to shorten the documentation "such as by using continuous text instead of separate document subsections for each ODD element".

With regard to the lifecycle, we may say that the ODD protocol covers foremost the aspect of model definition (i.e. *structure*). It does, however, not document the *process* of modelling, i.e. the theoretical knowledge involved (e.g. in the form of excavation reports and other scientific publications), the development of the model and the analysis of results. We argue that such a wholistic view is a good addition to ODD, since questions about choice of methods (why agent-based?), parametrisation (why are certain parameters included?) and so on can aid to reproduce not only model, but also get a view on its context. In this light, we might say that the documentation of the development inherits not only project-specific details such as deliverables and milestones, but also the documentation of all development steps taken. As already stated in Sect. 4.3, the development process of a M&S study happens in an iterative manner; it is important that all versions of this process or at least all changes from one to the next version of a model are documented as clearly as possible. Applied validation and verification methods (see Sect. 4.6) should be included as well—especially when dealing with informal techniques involving the subjective judgement (since that may state what parts of a model are ok and which need to be redefined in the next cycle).

4.5.2 Source Code

Source code has two great advantages: (1.) It *is already there*, i.e. there is little overhead in making it "documentation-ready", and (2.) it is *automatically kept synchronous to the model* (as opposed to every other form of documentation, which needs to be updated to reflect changes). The term "documentation-ready" does not refer to commenting, as comments get outdated when the model (and thus: the code) changes. Rather, we argue that *the source itself is a form of documentation*, if written narratively. Some aspects of this endeavour (also refer to Martin 2008) are:

Choosing proper names: Variables, functions, classes and packages should immediately reveal their purpose through their names.

Do One Thing Principle: Functions and classes should do one thing only (single responsibility), as expressed by their name. Whenever this does not hold, the model developer needs to split and recompose them (thus imposing a hierarchical structure where code one a higher-level defers lower-level responsibilities to subclasses or functions implementing lower-level behaviour).

Don't Repeat Yourself Principle: Generalising pieces of code that are similar, using concepts from Object Oriented Programming (i.e. inheritance, interfaces, etc.), makes code easier to read, extend and fix.

When adhering to these concepts, code published online (using for example www. openabm.org) can be understood by a large community of researchers, bypassing the need for secondary literature about the inner mechanics of a model to a certain extent.

4.5.3 Visualisation

Besides textual documentation, visualisation is crucial when trying document validate simulation models and thus make them "reproducible". In contrast to source code, which is the most immediate way to communicate the implemented model, the *process of modelling* and *fundamental structural ideas concerning the model* can often be better presented via visual concepts.

For a long time, visualisation was restricted to a very small range of possible applications in modelling and simulation: "[. . .] visualisation was limited to static images until the end of the twentieth century. This explains why visualisation was—and mostly still is—used as a post-processing step with the sole purpose of presenting results. On the other hand, increasingly powerful computers and display devices permitted to move from static images to (interactive) visualisation which has become a field of research in its own right" (Piringer 2012, p.1).[6]

Various parts described in this chapter, like the lifecycle itself, parameter-(input) and output definition and analysis or the later-described validation can be visualised. More specifically, one may depict (1.) the model structure, behaviour of agents and so on, typically using vector drawings for rendering graphs and diagrams (as for example in the Unified Modelling Language, see Booch et al. 2005). One may (2.) also visualise a simulation's runtime state (raster graphics showing agents and cells, see Kornhauser et al. 2009 for an extensive treatment on that subject) and to show its results; this is especially helpful in validation, as one can show runtime effects of the model (see Sect. 4.6). In addition, data used for the process of parametrisation can (or should) be analysed via "visual analytics", a new field which seeks to get "insight, not numbers" (Hamming 1962, in Preface) by employing a variety of depictive techniques. And (3.), quite importantly, one may also communicate the development process of a model *visually*.

Let us give a short glimpse of the possibilities that visualisation offers in the context of reproducibility (concepts are mentioned in the order as they appear in the M&S lifecycle):

- *Data Analysis*: Visualisation can support archaeologists and modelling experts in their collaboration. Visual analytics helps to analyse basic data, in order to support the detection of relationships and structural errors in the data. More precisely, "visual analytics combines automated analysis techniques with interactive visualisations for an effective understanding, reasoning and decision making on the basis of very large and complex datasets" (Keim et al. 2010, p.7). But visual analytics is not an end itself; it is the *employment of complex datasets* that makes the analysis of multivariate data (as base data) necessary. In that context, one needs to define, calibrate, and validate a model and should also represent the coverage of the parameter space adequately (think: depiction). Time-dependent datasets are special cases of complex datasets, which serve as longitudal base data (e.g. harvest yields, periods of illnesses) in the context of archaeology. Using visual analytics on that data supports hypothesis generation, intuitive integration of domain knowledge and also helps to make complex and unclean data nevertheless applicable.
- *Modelling Process*: The process of building models is highly iterative. Starting with data analysis and the generation hypotheses, it extends over the whole lifecycle and evolves in a spiral process. Research in *interactive model building*

---

[6]A number of aspects described in that report are used as a basis for this chapter.

(see Bertini and Lalanne 2009) aims to tightly integrate (1.) the specification and adaptation of models with (2.) an evaluation of results, in order to obtain well-understood models with less effort. Visualisation can aid in this process and, by doing so, also increase reproducibility. One specific example for that would be the visualisation of different parameterisations during the modelling phase: For example, reflections on inherent uncertainty in the parameters can be done when multiple simulation runs enable the modeller to sample the distributions of one or more parameters, in order to determine the uncertainty of the outputs (*sensitivity analysis*). A visual characterisation of both parameter and result space helps to identify all possible results that a particular model may possibly generate, and thus helps to simplify a model by omitting ineffective parameters. Furthermore, visualisation in the context of the modelling process can also help in calibration.

- *Structure*: Assessing visualisation approaches for representing a model's structure includes a high number of technologies, like visualisation of trees and graphs. Visualisation concepts for these models are well known. ABMs are a flexible, general approach. Thus, a detailed projection of those structures onto the model leads to a wide variety of model types, as agents can represent moving armies, vehicles on roads or something completely different. Different agents, different interactions, different rules and different structures make it almost impossible to provide a generally usable visualisation technique. Today, most visualisations of agent-based models focus on the model structure (especially the agents themselves), their behaviour and their interactions. Kornhauser et al. (2009) have proposed some guidelines for visualisation design specifically for agent-based modelling, which can be used to identify important model elements and help users to understand the model's behavior.

- *Runtime & Results*: Visual representations of simulation results include general-purpose statistical plots like bar charts and scatter plots (Tufte 1983) as well as diagrams addressing specific questions of the professional field. Techniques of visual analytics can be used for visualising (potentially very long) time-dependent data of single time series as well as a large number of time series of simulation results. Agent-based simulations constantly produce new data with every time step. A potentially large number of agents makes it difficult for the user to keep track of a particular agent or group of agents' position, colour, size, or shape. In this context, Healey and Enns discuss common visualisation tasks such as target and region tracking, boundary detection, and estimation and counting (Healey and Enns 2012). However, *visual summaries* of a simulation run are often far more effective for analysing the model and its implications than a look at individual agents. There are several examples over well-established techniques in that context (Tufte 1983, 1996, 2006), however, such "simple" depictive tools may get more complex once "drilling down" into the information. An example of such a "complex summary" could be a social network analysis reflecting the contact and interaction between agents.

Also, note that visual analysis can also be used for *excluding certain hypotheses* (rather than proving with reference to some data). In that sense, one can look at different scenarios (representing the hypotheses) and scrutinize the underlying data (visually), in order to save modelling efforts if a scenario can be easily rejected.

4.6 Verification and Validation

Verification answers whether a model is implemented correctly, i.e.

"Is the model developed right?"

Validation addresses the problem whether a model is able to answer a specific research question:

"Is the right model developed?"

Both can be seen as processes that happen in parallel to the development of the model and the simulation. Their aim is to guarantee a targeted development of the simulation study. Another important term connected to verification and validation is *credibility*. According to Law and McComas (2009), "a simulation model and its results have credibility if the decision-maker and other key project personnel accept them as correct." The higher the degree of scrutiny within the development process, the higher the possibility that the model is credible. But, "note that a credible model is not necessarily valid, and vice versa" (Law and McComas 2009, p.23).

In most cases verification and validation is carried out by the development team itself. The downside of this approach is that developers may tend to follow the same procedure that they use for development when they verify (i.e. they are caught in the same tracks). A better way is to conduct verification and validation by an independent third party (e.g. a "Verification and Validation Task Force", consisting of a mixture of people familiar with modelling and people connected to the field of study).

In the following, we give a description of verification and validation in respect to the generalized lifecycle (see Fig. 4.4).

- *Conceptual model validation* happens in the analysis and modelling phase. All abstractions, theories and assumptions are checked using mathematical and statistical methods.
- *Computerized model verification* deals with the substantiation of the right programming and implementation of the conceptual model. Different programming languages offer a broad variety of concepts to do this.
- *Operational validation* deals with the evaluation whether the model parameter is chosen right in respect to the purpose of the simulation. Here the biggest part of the validation takes place. It is important to remember that mistakes that are found in this part of the modelling lifecycle can be mistakes that were either made in the analysis and modelling part or in the computer programming and

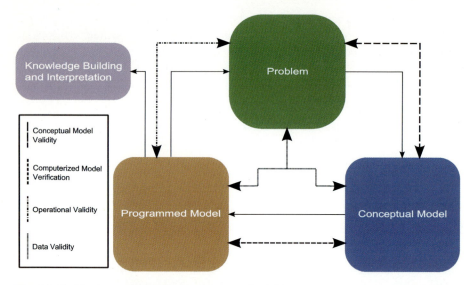

Fig. 4.4 Verification and validation in the generalized lifecycle of a simulation study based on (Sargent 2010)

implementation. The best way to do operational validation is the comparison to the real world problem. If it is not possible to do this, a comparison with other models should be done.

- *Data validation* helps to determine whether the data is correct. If assumptions are made to use the data, it helps to determine whether these are proper assumptions.

Arguably, the most simple validation technique is the *plausibility check*: It is an evaluation over whether the produced output is comprehensible, if the parameters are known. This can be done by either comparing the results to reality or, if that is possible, to other simulations. Another possibility in that context is to use expert knowledge. However, there are many more techniques like this, as is shown in due course. According to Balci (1994), these can be classified into six groups:

Informal Techniques: Techniques that rely on subjective judgement. This does, however, not mean that there is a lack of structure or formal guidelines being used in this case. Some examples also applicable to agent-based simulation are:

- *Audits*: A single person evaluates whether the simulation study is done in a satisfying way (e.g. checking whether it meets the preset requirements). As a byproduct, errors and mistakes may be uncovered.
- *Inspections, Reviews, Walkthroughs*: Each of these methods involves a group of people which is trying to evaluate whether the development is done in a satisfying way. These methods aim at different parts of the modelling lifecycle and use a variety of instruments.

- *Turing Test*: This testing technique was invented in 1950 by Alan Turing: A computer program is said to succeed the Turing Test if a real person cannot figure out whether the results are produced by a program (in this case: the simulation) or the results are taken from the real system.
- *Face Validation*: An expert checks whether results are reliable and reasonable. Results of other validation and verification strategies might support the expert in his decision.
- *Desk Checking*: The work of each team member is checked for correctness, completeness and consistency. As a side-note, model developers should not do desk checking by themselves, because mistakes are often overlooked or disregarded.

Static Techniques: The source code (which does not need to be executed, mental execution is enough) is automatically analysed by a compiler. To give some examples, *consistency checking* is a pre-step to compilation that can ensure a common programming style among the developing team. *Syntax analysis* is the basic compilation task can find wrongly spelled or grammatically wrong language constructs. *Data flow analysis* can furthermore detect undeclared or unused variables.

Dynamic Techniques execute the model and analyse its resulting behaviour:

- *Sensitivity analysis* analyses the impact of parameter changes on the output, by comparing multiple simulation runs in which parameter values have been systematically changed. An expert then has to check the results of this comparison. For example, a model that is very sensitive to input changes might easily lead to wrong results (or accumulated effects thereof), if the input data and parameters are not well-controlled.
- *Black-box Testing*: Input parameters are fed into the model and the accuracy of the output is evaluated without looking what happens inside the model.
- *White-box Testing*: This is more a verification technique than a validation technique, which we mention here for the sake of completeness. It is the same as Black-box Testing, albeit with access to the underlying code. Because this is known, a test run can demand for example that the whole code needs to be covered by a specific test. Or, it could demand that as many different logical conditions in the code are to be covered as many times as possible, thus substantiating the accurate operation of a simulation.
- *Bottom-up Testing*: This validation strategy is possible for simulations that are developed in a bottom-up manner: After sub-models are developed, an integration test is performed: This starts with the most basic functionality (e.g. simple movement of an agent), goes on to aggregate level (e.g. steering behaviour of an agent) and continues onto the topmost level (e.g. movement of a crowd).
- *Top-down Testing*: This validation strategy is possible for simulations that are developed in a top-down manner. It is the counter-part of bottom-up testing: After a top model is completely developed, a hierarchical integration test for the sub-models is performed.

- *Graphical Comparisons*: Real-world data is compared to the modelling variables and parameters, assisted by graphs in which e.g. linearity, exponential or logarithmical rise, periodicities and so on are analyzed. Such comparisons can furthermore be used not only for checking against real-world data, but also against other models.
- *Visualisation*: The simulation output is visualised through the whole simulation run. This technique can only be seen as validation strategy if the results are checked for plausibility afterwards.
- *Predictive Validation*: The model is fed with known (past) input parameters. The model output is then compared to the real world behaviour. Of course, this can only work when real-world data (both input and output) is at hand.
- *Statistical Validation*: A model could be statistic validated when the system is completely observable. For example, variance analysis, confidence intervals and so on may be done for the output. Whether this is a good validation strategy or not depends on the inner mechanics of the model. It is thus important to use many different data sets to see whether the statistical statements deduced are always the same, and the technique can therefore be applied.

Symbolic Techniques are for checking *cause-effect relationships* (through graphs), checking functional correctness of each part of the simulation (this is also called *partition testing*), testing specific portions of a simulation by supplying it generated test data that should cause a certain execution (*path analysis*), and lastly *symbolic evaluation* that tests all evaluated paths in a program to find out if some are unnecessary or uncorrect. In later work (Balci 1997), this category was incorporated into the dynamic and static techniques.

Constraint-Based Techniques: Supported by constraints, model correctness is warranted. These constraints can be seen e.g. as run-time assertions, which ensure that the simulation stays in a well-defined state. Balci (1997) gives this category up completely, and incorporates these concepts into the other categories.

Formal Techniques: Mathematical deduction and other formalisms are used to e.g. check the correctness of the simulation. However, as Balci (1994) admits, "Current state-of-the-art formal proof of correctness techniques are simply not capable of being applied to even a reasonably complex simulation model" (Balci 1994, p.218). Especially for agent-based models, formal techniques are still in a very early stage, and a general formal approach (for all scenarii where agent-based models are used) is still beyond the current state of the art.

4.7 Verification and Validation Specifically Targeted at Agent-Based Models

Agent-based models are characterized by a complex behaviour that cannot be entirely observed. This is why informal techniques can always be used, because these are subjective methods. Methods with a very formal background, as already stated by Balci (1994), cannot be used for the entire model, due to high amount of complexity involved. But even when some parts of an agent-based model can be formally validated, this does not imply that the composition of all parts is also valid.

One specifically "agent-based" validation strategy is called *Immersive Face Validation* (Louloudi and Klügl 2012). It is a strategy in which one tries to see the model *as from an agent's point of view*. Thus, the path of an agent is tracked through the whole simulation. It is then analysed and checked against the model's assumptions, using a virtual environment (i.e. 3D visualisation) in which the behaviour of one particular agent can be assessed. As the authors state, "after completing one full simulation run, the evaluator should describe the overall comments briefly in a post processing activity and reply on model-specific questionnaires" (Louloudi and Klügl 2012, p. 1257). According to the categories described earlier, this is thus an informal technique.

A procedure for the validation of an agent-based model, based on Klügl (2008), is shown in Fig. 4.5: The authors assume that the model has already been subjected to a conceptual model validation (see Sect. 4.6) and is now ready for execution ("Run-able"). This execution is assessed using a face validation, leading to either a sufficiently plausible model or a re-iteration ("not sufficiently valid"). The plausible model then undergoes a sensitivity analysis, which leads to an assessment regarding which parameters are influential and which ones are without effect. The latter ones can be deleted (together with code parts connected to them), leading to a "Minimal Model". Then comes a calibration part, in which parameters have to be set such that the model produces valid results. Note that the calibration process itself is not

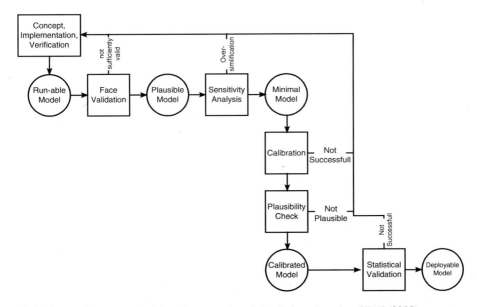

Fig. 4.5 A validation methodology for agent-based simulations, based on Klügl (2008)

a validation method—it is merely a part of that procedure. A following plausibility check is then conducted, which is "basically the same as the previously discussed face validation, yet may not be executed as intensively as before as we assume only limited changes in the simulation outcome" (Klügl 2008, p. 43).

To end with, we would also like to mention the work of Muaz et al. (2009), who have introduced an approach called VOMAS (Virtual Overlay Multi Agent System): Special agents—so called VOAgents, monitor the agents of the simulation ("SimAgents"). This VOAgents can get simple rules which tell them what to do, for example rules that tell them to trace single SimAgents, log SimAgents with certain attributes or variable values and so on. Also, groups of SimAgents with certain attributes or variable values can be monitored in that manner. Basically it is an extended and individualized protocol module with a high degree of flexibility, if implemented right. This system can be used to support the validation of agent-based models *within the same environment*. Thus, care has to be taken that VOAgents are really only "external observers" and not influencing the model's state.

4.8   Conclusion

One possible purpose for employing any kind of simulation is the possibility to test one or more hypotheses. The first thing that comes to mind when thinking about a successful simulation study is that the simulation has to have an outcome that helps to confirm these (i.e. an *expected* outcome). But simulations can also be used to gain more knowledge about the system under study: It can prove whether given assumptions or abstractions might be wrong; by *falsification*, it helps to reduce the number of possibilities to the ones that seem reasonable. However, falsification may also be hard to achieve: "If the model contains too many degrees of freedom, an automatic optimizing calibration procedure will always be able to fit the model to the data" (Klügl 2008, p. 41). As example for a model in which falsification was expected but not manageable, the author names the EOS project (Doran and Palmer 1995): "Two competing plausible hypotheses for explaining the emergence of social complexity were simulated, both were able to reproduced [sic!] the observed emergent structures. Thus, no theory was rejected, the modeler had to admit, that without additional data no discrimination between the two hypotheses could be made" (Klügl 2008, p. 41). An exploration of the subject of falsification is an interesting field for further investigation (e.g. in Popper in press).

We argue that in many cases there is no single point at which a model—especially an agent-based one, can be made "reproducible", "validatable" or "falsifiable" (all of these terms are *properties* and not tasks to conduct). Instead, we have to look at the whole lifecycle of an M&S study and address aspects such as documentation, verification and validation, which were presented herein. It is also clear that these need to be carried out throughout the entire development process. Moreover, as initially stated, the knowledge produced in such manner should be *transparent*, i.e. openly available for professional discourse and scrutiny.

References

Balci O (1994) Validation, verification, and testing techniques throughout the life cycle of a simulation study. Ann Oper Res 53(1):121–173. DOI 10.1007/BF02136828, URL http://link.springer.com/10.1007/BF02136828

Balci O (1997) Verification, Validation and Accreditation of Simulation Models. In: Proceedings of 1997 Winter Simulation Conference, Atlanta, GA, USA, pp 135–141

Bertini E, Lalanne D (2009) Investigating and reflecting on the integration of automatic data analysis and visualisation in knowledge discovery. SIGKDD Explor 11:9–18

Booch G, Rumbaugh J, Jacobson I (2005) The Unified Modeling Language User Guide, 2nd edn. Object Technology Series. Addison-Wesley Professional

Doran JE, Palmer M (1995) The EOS Project: Integrating Two Models of Palaeolithic Social Change. In: Gilbert N, Conte R (eds) Artificial Societies: The Computer Simulation of Social Life. UCL Press, London, pp 103–125

Freire J, Bonnet P, Shasha D (2012) Computational Reproducibility: State-of-the-Art, Challenges, and Database Research Opportunities. In: Proceedings of the 2012 ACM SIGMOD International Conference on Management of Data, pp 593–596

Grimm V, Berger U, Bastiansen F, Eliassen S, Ginot V, Giske J, Goss-Custard J, Grand T, Heinz SK, Huse G, Huth A, Jepsen JU, Jørgensen C, Mooij WM, Müller B, Pe'er G, Piou C, Railsback SF, Robbins AM, Robbins MM, Rossmanith E, Rüger N, Strand E, Souissi S, Stillman RA, Vabø R, Visser U, DeAngelis DL (2006) A standard protocol for describing individual-based and agent-based models. Ecol Model 198(1–2):115–126

Grimm V, Berger U, DeAngelis DL, Polhill JG, Giske J, Railsback SF (2010) The ODD protocol: A review and first update. Ecol Model 221(23):2760–2768

Hamming R (1962) Numerical Methods for Scientists and Engineers, 1st edn. McGraw-Hill, New York

Healey C, Enns J (2012) Attention and visual memory in visualization and computer graphics. IEEE Trans Visual Comput Graph 18(7):1170–1188

Keim D, Kohlhammer J, Ellis G, Mansmann F (eds) (2010) Mastering the Information Age – Solving Problems with Visual Analytics. Eurographics Association, URL http://www.vismaster.eu/wp-content/uploads/2010/11/VisMaster-book-lowres.pdf

Klügl F (2008) A Validation Methodology for Agent-Based Simulations. In: Proceedings of the 2008 ACM Symposium on Applied Computing, ACM Press, New York, NY, pp 39–43. DOI 10.1145/1363686.1363696, URL http://portal.acm.org/citation.cfm?doid=1363686.1363696

Kornhauser D, Wilensky U, Rand W (2009) Design guidelines for agent based model visualization. J Artif Soc Soc Simulat 12(2):1. URL http://jasss.soc.surrey.ac.uk/12/2/1.html

Law AM, McComas MG (2009) How to Build Valid and Credible Simulation Models. In: Rossetti MD, Hill RR, Johansson B, Dunkin A, Ingalls RG (eds) Proceedings of the 2009 Winter Simulation Conference, Austin, TX, USA, pp 24–33

Louloudi A, Klügl F (2012) Immersive Face Validation: A New Validation Technique for Agent-Based Simulation. In: Proceedings. 2012, IEEE, Wroclaw, Poland, pp 1255–1260

Martin RC (2008) Clean Code: A Handbook of Agile Software Craftsmanship. Prentice Hall, Upper Saddle Rive

Muaz AN, Hussain A, Kolberg M (2009) Verification & Validation of Agent Based Simulations Using the VOMAS (Virtual Overlay Multi-Agent System) Approach. In: Proceedings of the Second Multi-Agent Logics, Languages, and Organisations Federated Workshops, Torino, Italy

Ogden CK (1940) General Basic English Dictionary. Evans Brothers Limited, London. ISBN: 0874713625

Ogden CK (1968) Basic English: International Second Language. Harcourt, Brace & World, New York. URL http://ogden.basic-english.org/isl.html

Piringer H (2012) A Survey and Assessment of Visualization Approaches in the Context of IFEDH. Tech. Rep., VRVis, URL http://ifedh.dwh.at/sites/default/files/IFEDH_Visualization_Survey.pdf

Polhill JG (2010) ODD updated. J Artif Soc Soc Simulat 13(4):9. URL http://jasss.soc.surrey.ac. uk/13/4/9.html

Popper N (in press) Comparative Modelling & Simulation. PhD Thesis, Vienna University of Technology

Sargent R (2010) Verification and Validation of Simulation Models. In: Proceedings of the 2010 Winter Simulation Conference, Baltimore, MD, pp 166–183

Stodden VC (2010) Reproducible research: addressing the need for data and code sharing in computational science. Comput Sci Eng 12(5):8–12

Tufte E (1983) The Visual Display of Quantitative Information. Graphic Press, Cheshire

Tufte E (1996) Visual Explanation. Graphic Press, Cheschire

Tufte E (2006) Beautiful Evidence. Graphic Press, Cheshire

Chapter 5
Geosimulation: Modeling Spatial Processes

Andreas Koch

5.1 Introduction

One of the most stimulating epistemological reflections in dealing with reality contemporarily is the fact that we are gauging the world with the images we have in our minds or we have produced as pictures, graphs, photographs, and (cartographic) maps—and not vice versa. A shift from imagining to imaging has been taken place in science and Lebenswelt (Belting 2005, p. 24). The power of imagination derives from images we create utilizing different devices and techniques, methods and rules. The challenges we are confronted with are readability, reliability, transferability, and self-efficacy with regard to the patterns, relations, contexts, and identification we perceive sensually, mentally, and emotionally.

This is especially true in the field of modeling and simulation in general, and geosimulation in particular, with their intuitively captivating illustration of different kinds of processes generating, perpetuating, or changing physical and societal structures in geographical spaces. This allows for a different kind of reasoning about phenomena that surround us as Resnick (1997, p. 49) points out with respect to the application of computational simulation tools: "In short, I am more interested in stimulation than in simulation". Another implicit implication of the changing relationship between imagination and image is the understanding about how we are accessing the world, reality, or the truth. The notion of image points to the fact that this access is always achieved through translations done by intermediaries. Our references to the world are thus permanently mediated and models are one of the most prominent and basic intermediaries.

A. Koch (✉)
University of Salzburg, Salzburg, Austria
e-mail: andreas.koch@sbg.ac.at

© Springer International Publishing Switzerland 2015
G. Wurzer et al. (eds.), *Agent-based Modeling and Simulation in Archaeology*,
Advances in Geographic Information Science, DOI 10.1007/978-3-319-00008-4_5

Models are defined by three characteristics (Stachowiak 1973): (1) representation; (2) simplification; (3) pragmatism; these in turn confirm the model approach by a self-referential application of the model idea to the model definition. Every model is a representation of a natural or artificial original.

1. The original is not equal to reality but equal to another model. A model does not cover every attribute of the original, but only those whose properties are of any relevance for the modeler.
2. The original is a more or less conscious setting of selected objectives and not a holistic entity. A model is not in itself a representation of the original.
3. One always has to ask "why", "for whom", "for what" a specific model has been created in order to specify the intention of that model. In other words, there is always a close relationship between a model and the modeler's style of reasoning.

The latter epistemology appears to be appreciated as well by archaeologists applying computational approaches to model and simulate ancient socio-spatial phenomena. Lake (2010, p. 12), for instance, by referring to Mithen (1994), ascertains "the intention of the researcher(s) conducting the simulation" as one purpose of using simulation in archaeology. Accordingly, the seemingly distinctive perspectives of a "faulty understanding of the reality we are modeling or [a] faulty modeling of the reality we are seeking to understand" (Costopoulos et al. 2010, p. 2) do in fact coalesce. Putting all three characteristics together, it is not only obvious that the map is not the landscape, but also that the landscape is always a represented, recognized composition. "There is no original [in the sense of reality; A.K.] from which to copy. Yet the end-result is image-like; it is a gestalted pattern which is recognizable, although it is a constructed image" (Ihde 2006, p. 84).

Recognizing and accepting this epistemological and methodological frame implies a couple of empirical and theoretical potentials in terms of gaining new, different and stimulating insights without referring to an absolute reality but enabling a continuous movement of translation between models. Geosimulation in this respect is a spatially and temporarily scale-sensitive observational instrument, making processes and their causes and conditions visible. Though notions of society, space, and time are commonly used both in science and ordinary communication, they are highly abstract due to their intangible nature. Utilizing instruments of observation is not new, but their meaning and mission has changed accordingly. "Whereas telescopes and microscopes render phenomena visible by affecting the scale of 'tangible' entities through optical processes of resolution, simulation renders 'visible' the effects of parameters and forces such as time, dynamic interactions, [. . .]. Thus, simulation, by constructing images, may translate absolutely nonvisual events into a visual media!" (Küppers et al. 2006, p. 8). In this paper then, modeling spatial processes is an attempt to tie spatial intangibles with social intangibles across time by referring to their tangible counterparts.

There has been put much methodological effort to render abstract socio-spatial processes visible. In geosciences in general and archaeology in particular, attempts to geovisualize complex relationships between social and spatial interactions as well as among these are one of the core concerns in order to gaining advanced insights into pattern recognition, knowledge about so far unprecedented (cor-)relations or (statistical) comprehension of simulated results against empirical records. Demands and challenges towards visualization techniques thus remain a wishful and necessary task, most notably within the realm of geovisualization and geosimulation as Aldenderfer (2010, p. 54) convincingly claims from a data acquisition point of view: "Because of the masses of data that can be created by even a modest simulation, we are in serious need of tools of all kinds to help us "see" our results".

In the remainder some theoretical reflections on the relationships of "society and space", "scale and process", and, methodologically, "agent-based modeling and geographic information science" from a geographical-geosimulation angle will be discussed. This is also the prevailing perspective of a couple of theoretical debates in computational archeological literature (e.g. Costopoulos and Lake 2010; Kohler and van der Leeuw 2007).

5.2 Society and Space: Dealing with Intangible Assets and Infrastructures

Spatiality, temporality, and sociality are highly abstract and fuzzy notions due to their intangible nature. They are, simultaneously, core principles in creating order. Space enables ordering by means of juxtaposition; time involves the process of succession, and the social dimension contributes a basis for togetherness (Fischer and Wiegandt 2012; Sennett 2012; Tate and Atkinson 2001). All three dimensions are mutually and interdependently tied together, e.g. the composition of a place depends on the physical structure and the people who assign this place for its use according to their social meanings which may vary over time; togetherness implies that beyond social contiguity there is some kind of spatial proximity, be it geometrical (face-to-face) or topological (social network relationships).

A central challenge for these three layers, their respective independent properties and dynamics, as well as their interdependent relationships, is the quest for patterns and their underlying rules. Another approach is to investigate simple social interaction and local neighborhood rules in order to research the emerging complexity that potentially happens at the macro-scale. There is a large body of literature in the computational social sciences, whose methodological distinction is applying either Cellular Automata (CA) or Agent-Based Models (ABM) in a broad sense. Since it is almost impossible to survey and thus cite all the literature of this field, a subjective selection is presented in the following paragraphs.

A recent and extensive review with an explicit reference to Geographical Information Systems (GIS) is given by Heppenstall et al. (2012). Benenson and Torrens (2004) published an approach entitled "Geographic Automata Systems" which extends common ABM applications towards an explicit and independent implementation of space. Gimblett (2002) edited a volume which discusses potentials of integrating GIS and ABM for simulating social and ecological processes, while Kohler and Gumerman (2000) investigate the coupling of social and spatial processes utilizing ABM, including two contributions (Kohler et al. 2000; Dean et al. 2000) who analyze settlement development and cultural transformations of the prehistoric Anasazi population in today's federal state of Arizona, USA. The CA approach in general has been extensively analyzed by Wolfram (2002), in connection with geospatial topics, like spatial analysis techniques, mainly in urban contexts. Batty (2005) and Maguire et al. (2005) thoroughly discuss opportunities and pitfalls of the CA technique. A specific application example of using agent methodology to simulating visitor behavior in natural landscapes is presented by Gimblett and Skov-Petersen (2008).

Furthermore, seminal publications in these fields are, among others, Troitzsch et al. (1996) and Conte et al. (1997), focusing on social microsimulation, as well as Epstein and Axtell (1996) with their Sugarscape model representing an early bottom-up approach. A recent publication by Squazzoni (2012) introduces the field of agent-based computational sociology, and Epstein (2006) provides a generative social science with an explicit spatial integration if necessary. In addition, there are a couple of conference proceedings, early examples may be Sichman et al. (1998) or Moss and Davidsson (2001). One of the most widely known online journals in the field of agent-based social simulation is the Journal of Artificial Societies and Social Simulation (JASSS, see: http://jasss.soc.surrey.ac.uk/JASSS.html), furthermore services like the Open Agent Based Modeling Consortium (http://www.openabm.org/) or the GIS- and ABM-related blog (http://www.gisagents.org/) provide interested readers with additional information, open software tools or discussion boards. Computational archaeology can well be integrated into this frame of socio-spatial interrelatedness. Janssen (2010), for instance, developed an agent-based model of prehistoric societies embedded into an abstract but explicit spatial context in order to investigate interdependencies of demographic and climate variabilities against potential strategies of social adaptation and adaptability. A similar model purpose has been pursued by Berger et al. (2007) who are interested in community resilience from a socio-spatial functional perspective. They, however, refer to a specific geographical region, the Middle Rhône Valley in France. A CA-based approach for representing the spatial domain has been chosen by Smith and Choi (2007) to simulate the emergence of inequality in small communities. Lake (2000) developed a multi-agent model which equips agents with cognitive spatial maps by integrating a GIS into his MAGICAL software tool in order to derive individual and cultural learning.

All models dealing with socio-spatial phenomena have to begin with a decision about the smallest, indivisible unit. In the social world this may be an individual, a household, company or institution. The spatial world can be composed of points, lines, and polygons, representing cells or areas; spatial units may represent bus stations, parcels, buildings, political or statistical territories, catchment areas, or landscape entities to mention but a few. Both types of individual units (regular cells and irregular areas) can be represented by agents who are equipped with properties and exhibit some rule-based interactions to other agents. There are a huge range of human agents' properties, since agents can be autonomous in their collective behavior or decision making. Agents are heterogeneous with respect to attributes like normative attitudes, opinions, desires, and intentions, as well as age, family status, or income. They can be pro-active or reactive, communicative, mobile or capable of learning and adaptation (Crooks and Heppenstall 2012, p. 86ff; O'Sullivan et al. 2012, p. 114ff; Epstein 2006, pp. xvi–xviii). Nevertheless, the actual implementation of properties depends on the precise model purpose and usually does not emerge comprehensively. Spatial agents may vary in properties as well; for example, they map land-use, land-cover, real estate prices, developed or undeveloped parcels, symbolic or normative ascriptions.

The design and functionality of relations can be complex, since there are three distinct levels of interactions each composed of different quantities and qualities of relationships: (1) interactions among human agents, (2) interactions among locational agents, and (3) interactions between human and locational agents. Though depending on geographical and/or resolution scale and thus on the model purpose, "locational agents" are defined here as the smallest spatial units used in a geosimulation model and can be seen as the spatial counterparts of human agents in the social world. The possibility of conceptualizing a mutually effected social-spatial model framework of interacting individual units is one of the exceptional characteristics of agent-based geosimulation—or, as Epstein (2006, p. 5) puts it in a nutshell: "How could the decentralized local interactions of heterogeneous autonomous agents generate the given regularity?" Apart from the scale-dependent micro-macro link (see Sect. 5.3) a problem arises with heterogeneity and autonomy if they are set as absolute. In fact, a dynamic struggle with their complementary parts of homogeneity and dependency mirrors societal conditions more appropriately, leading to a dialectic synthesis of inter-scaling and inter-dependent processes.

The current state of a human agent, as well as its variation and alteration, is a function of time, spatial, and social conditions. Though a few of these states can be assigned as intrinsic, like age and health status, most of them rely on exogenous influences. Variables like income, family status, housing situation, places of living, working, and recreation, social positions and roles, membership, professional status, and social embeddedness, all refer to a relationship between the single unit and the multi-leveled superior environments. The term "social" indicates a reference to larger entities as communities or societies. Modeling and simulating human agents' actions and behaviors thus needs to incorporate a kind of "social auto-correlation" which reveals social distances among the members of a community or society, representing the interplay of single units and the respective collective. This in turn enables conclusions to be drawn about social phenomena of lifestyles, ethnicities, migration backgrounds, normative and religious attitudes, to mention but a few. ABM methodologies have developed different conceptual frameworks in order to operationalize "social auto-correlation" (see, for example, the review of Kennedy 2012). One approach refers to beliefs, desires, and intentions (BDI; Rao and Georgeff 1995) which contextualize agents' knowledge, perceptions, motivations, and attitudes with social-environmental facts, based on a social constructivist epistemology (for a current review see Searle 2010, who primarily draws on status functions, collective intentionality, and institutional facts as explanatory objectives to derive the macro-level of society). A similar framework is based on physical, emotional, cognitive, and social factors (PECS; Schmidt 2002). "This framework includes a representation of the human mind, specifically perception and behaviours, and mathematical representations of physiology, emotion, cognition, and social status" (Kennedy 2012, p. 175). An approach which predominantly considers human decisions against environmental conditions by analyzing corresponding data is called "fast and frugal" (Gigerenzer 2007). All cases exhibit an awareness of a necessary multi-dimensional and multi-scaled coupling of social entities which is theoretically reflected by, among others, Giddens (1990) and Latour (2005).

This description of a human agent can be translated to a single locational (or place) agent as well. This type of agent may also have some intrinsic properties, for example, soil quality, slope, and land-cover; of at least equal significance is, however, the relational context to other locational agents, i.e. the neighborhood effect (whereby neighborhood is understood here in a broad sense, encompassing geometrical and topological as well as semantic concepts of nearness). Aldenderfer (2010, p. 61) points to this aspect of spatial proximity when advocating true geosimulation approaches as beneficial for archaeological investigations because of their georeferencing capability of social and spatial objects and thus their ability to relax spatial relations to noncontiguous neighbors. This capability restricts Cellular Automata approaches with their inherent concept of neighborhood as being contiguous and cell-like. Housing prices, location and allocation of infrastructures and services, and land-use conflicts all depend crucially on the spatial configuration of single place units. In this regard, spatial auto-correlation is also a meaningful method. Geocomputation and geospatial analysis provide a huge range of sophisticated techniques which aim at investigating (geo-)statistically distance-weighted and directional variations of geo-referenced objects or events. It is beyond the scope of this contribution to give an extensive review of all the available tools in geostatistics and geographic data-mining, thus a brief and selected overview is being presented here (for an introduction to geostatistics see, among others, Leuangthon et al. 2008; Remy et al. 2009; for an introduction to data-mining see, for example, Han and Kamber 2006; Miller and Han 2009).

One approach is point pattern analysis (PPA) which compares an empirical distribution of points representing geo-referenced events with a theoretical (random) distribution (Lloyd 2007, p. 171ff; Wang 2006, p. 36ff). The statistical aim is to examine whether or not the empirical distribution differs from the theoretical one, either representing a (significant) tendency towards clustered or towards regular distributions. PPA is appropriate when investigating settlement structures or archaeological sites in order to gain a better understanding of the spatial (and spatio-temporal) influences of community or social development. Another approach is deterministic interpolation. In general, interpolation attempts to deduce knowledge about a study area by estimating data values for any arbitrary location within the study area by referring to empirical data measured at specific sample points (Johnston et al. 2001). The technique of "referring to" is the sum of a pairwise calculation of the variance between the point of interest (POI; it represents an unknown value) and all empirical points, inversely weighted with the distance between each two points (the POI and one empirical point). In other words, spatial auto-correlation is being incorporated assuming that the closer two points are located, the more similar they are with respect to the data value. Inverse Distance Weighted (IDW) interpolation is one of the most widely used techniques of a deterministic interpolation. This method can be extended by integrating the spatial auto-correlation among all empirical points in order to achieve additional spatial information when estimating an unknown data value within the study area. This alternative is referred to as probabilistic interpolation and Kriging techniques are the most well-known in this field of geostatistics (Remy et al. 2009).

For any agent-based model dealing with socio-spatial simulation it can be concluded that individual human behavior is neither completely self-determined (which would be synonymous to probabilistic and unpredictable in social contexts) nor completely socio-spatially and socio-culturally determined (and thus synonymous to predictable and calculable). The same applies to individual places. A common challenge for agent-based modeling is to choose the appropriate level of generalization, according to the model purpose, the theoretical background, and the available data. Interdependencies between human actions and social norms, between place properties and spatial structures, and between the two realms are constantly floating and fuzzy when observed—and including dynamic temporal processes further complexifies these interdependencies. Furthermore, socio-spatial agent-based modeling deals with intangible assets and infrastructures like beliefs, intentions, togetherness, and neighborhood, trying to visualize these by using "solid" parameters. The endeavor of recognizing spatial patterns in community life and social patterns in spatial structures implies consideration of the betweenness of the poles, heterogeneity-homogeneity/local-global, instead of the poles themselves.

Spatial analysis in general and geostatistics in particular appear to be promising tools in archaeological geosimulation as well. According to Lake (2010, p. 12ff), the purposes of computer simulations in archaeology derive from empirical hypothesis testing, theoretical reasoning on structures, processes, and functions, and from methodological explorations into statistical inferences. Since all three levels of scientific endeavor are capable of dealing with (and do refer to) complex spatio-temporal dynamics, the use of geosimulation techniques outperforms simpler CA approaches by releasing patterns of agents' actions socially (with respect to collective vs. individual action framing) and spatially (with respect to local vs. global spatial framing).

5.3 Scale and Process: Taking Relations and Interdependencies into Consideration

According to Benenson and Torrens (2004, p. 25ff), the human-society and place-space link described in the previous section can be formalized in agent-based geosimulation models at a general and coarse level as follows:

$$G \sim (K; S, T_s; L, M_L; N, R_N),$$

with G = geosimulation model
   K = ontologies (e.g., polygon or raster representation of space, agents representing humans or households)
   S = set of human and locational agents' states (e.g., income or family status in the former case, housing prices or land-cover types in the latter case)
   T_S = transition rules of S; in discrete time steps agents change their states according to their internal characteristics and to their social-spatial environment conditions, thus K, S, L, and N
   L = location; geo-referenced specification of human and locational agents
   M_L = movement rules of mobile human agents; they depend on K, S, L, N, resolution, and level of abstraction
   N = neighborhood; it encompasses (depending on the model purpose) the spatial and social neighborhood
   R_N = neighborhood rules; they define the criteria of spatial relations (geometrical and/or topological, adjacency and/or distant relations, vision of agents, etc.) and refer as well to K, S, L, and N

This kind of commonly used agent-based model represents the bottom-up type of simulations which implement social and spatial domains as latent variables of individual human and locational agents. The overall aim is to deduce social and spatial macro-structures from human and local decisions and micro-characteristics, respectively. Arguably the most commonly used concept in urban social-spatial agent-based modeling is Schelling's residential segregation approach (Schelling 1969, 1971) which has gained much attention (see, for example, Bruch 2006; Crooks 2008, 2012; Fossett and Senft 2004; Fossett and Waren 2005; Koch 2009; Laurie and Jaggi 2003; Pancs and Vriend 2007; Resnick 1997). The fascinating and stimulating issue of this model, whose theoretical and methodological foundations can be translated into different societal contexts across historical eras, is this inductive-driven phenomenon of emergence (Holland 1998; Johnson 2001), i.e. that the macro pattern of social-spatial structure cannot be derived from the micro pattern of human motives. The benefit of this type of simulation model is its focus on the processes taking place subtly and gradually. The noticeable and visible result of residential segregation usually allows for a retrospective speculation but does not derive sufficient information about the development. "Schelling's model is excellent because it distils the key features enabling us to understand how segregation might arise. The model does not presume to tell us about the entire working of the social and economic world, but focuses on the task at hand, namely to explain why weak individual preferences are consistent with strong and persistent patterns of segregation" (Crooks 2012, p. 369). Figure 5.1 illustrates a potential outcome of a Schelling-styled segregation process with a preference for human agents of at least 30 % neighborhood identity.

   Though the model provides, according to its purpose, many scientific insights about the process itself and the phenomenon of emergence, it incorporates only the individual and local scale in order to simulate urban social-spatial dynamics and changes. The model can only succeed in doing so by presupposing a couple

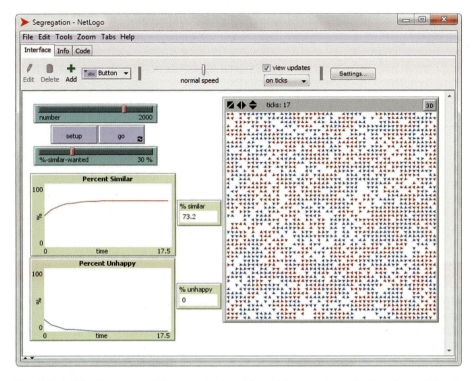

Fig. 5.1 A Schelling residential segregation model. Initially a random-distributed population is assumed with one agent property and two manifestations (*blue* and *red*). The threshold value of neighborhood satisfaction is set to 30 %, below this value agents are forced to move to the next unoccupied cell. After a few time steps the proportion of "unhappy agents" has been reduced to 0 %. *Source*: NetLogo Model Library, see Wilensky (1999)

of preliminaries which do not seem to be necessary, neither with respect to the phenomenon of interest, nor with respect to the empirical original the model is referring to. The following remarks are explicitly related to the problem of social and spatial scaling and do not devalue the model benefits as such:

1. The model takes a randomly distributed population as a starting point which seems empirically unrealistic and epistemologically avoidable. Even if empirical data with spatially high resolution and historically sufficient reliability are missing, it would be more appropriate to assume non-random distributions of population. This can be stated for ancient and medieval societies where guilds and status groups had lived spatially close together, as well as for modern societies with their socially distinct housing areas in urban districts or suburban regions.
2. Furthermore, all agents behave the same way. There are neither within-community nor inter-community differences.

3. It is assumed that social facts as norms, rules, cultural artifacts, and laws mean the same for all agents and are interpreted in the same way which in turn induces identical behavior.
4. There is only one characteristic on which decisions are based; social change is thus excluded, and it is not only the individual decision process which is modeled in a deterministic style with regard to the macro-scale result, but also the macro-scale result as well, since the model stops if all micro-motives are satisfied.
5. In addition, there is one single decision rule and one single threshold value which determines agents' actions. In other words, (residential) satisfaction is reduced to one dimension with one discrete reference measure for all agents, which reduces their scopes of decision-making to pure reaction. Migration should not be considered as the solitary solution to avoid dissatisfaction with a location, because it is a socially far-reaching and expensive decision in everyone's life.

Though depending on the model purpose, it is worth thinking about a more suitable adaptation of agents' behavior in terms of differentiation and flexibility. The aim could be a model approach where social and spatial agents exhibit diverse (individualized) actions within a social range (collectively shared norms or attitudes) changing over time and across space. The well-known archaeological Anasazi simulation model (Dean et al. 2000), for instance, diversifies agents' characteristics but does not use historical settlement locations. The inequality model of ancient societies or communities (Smith and Choi 2007) varies agents' demographic and economic characteristics but allow only two different behaviors (cooperation and defection) which will be inherited by offspring. Lake's (2000) approach, on the other hand, applies a strict individualized agent setting, and criticizing homogenous collective decision making in hunter-gatherer archaeological simulations: "These archaeological studies have not, however, harnessed the full potential of multiagent modeling. [...] In contrast, the MAGICAL software allows each individual to behave according to a potentially unique set of principles, which means that it is possible to simulate individuals thinking and behaving differently according to factors such as age, gender, and social standing." (Lake 2000, p. 109).

From a spatial perspective the problem with a Schelling-styled simulation model is that the neighborhood is conceptualized exclusively geometrically, i.e. space is perceived as being continuously given, without blank spaces, and being evenly important (at least all eight adjacent neighbor cells of a given location, according to Moore's neighborhood; see Iltanen 2012, p. 73; Patel et al. 2012). A topological conceptualization is missing, which takes social relations as networks into account. Furthermore, it is assumed that local knowledge about neighbors is total and immediately updated after each time step which suggests an unrealistic imagination of complete rationality. People may, in addition, migrate, but their social relations with former neighbors may remain significant and reliable, while the relations with new and close neighbors are rudimentary. Finally, both ubiquitous neighborhood and social network relations are unlikely to remain static and stable over time even if no migration takes place. Another spatially and socially influencing force is urban planning and social neighborhood management. These institutions are engaged in avoiding ghettoization, gentrification, gated communities and other forms of social segregation. Social housing and social capital empowerment projects with external partners aim at maintaining vivid and diversified local communities. This in turn could lead to some persistent strategies of individuals or households, although their dissatisfaction level has been increased.

In conclusion, agent-based residential segregation models tend to concentrate on the individual and local scale leaving aside the complementary scales towards society and space. There is much to be said for incorporating macro-scale domains into geosimulation models. Over the entire history of societies there have been social structures and normative rules available which provided for collective order. Ancient and traditional societies had been stratified and segmented by social origin, class, profession, gender, kinship, and local context. Contemporary western societies are characterized as functionally differentiated systems, embedded in globalized and re-localized structures, and today it is education, personal skills, flexibility, financial resources, and technological capabilities that promote social and spatial mobility. Social change towards modernization is highly paradoxical (see Fig. 5.2) making the modeling and simulation of socio-spatial phenomena a complex endeavor.

"Structural differentiation is today an all-embracing phenomenon which increasingly is determined by economic forces of efficiency and optimization. [. . .] Not only has the spatial and social scaling changed but we also maintain both intimate and distant relationships, face-to-face and virtual, without any specific correlation between distance and emotional nearness" (Koch 2012, p. 10f). The cultural sphere is characterized by a struggle between pluralization and generalization: "On the one hand we are aware of a plenitude of lifestyles, family constellations, and educational institutions which cultivate their own values and norms and by so doing establish specific mechanisms of access. On the other hand a global homogenisation of taste and preference in sports, fashion, music or literature can be recognised. Distinction, thus, is relative to interpret" (Koch 2012, p. 11). Individualization is in constant flux between increased autonomy and dependency. "Dependent autonomy or autonomous dependency extends scopes of acting because one can delegate and/or integrate tasks individually based on personal needs. Mutual out- and

down-scaling	pluralization
DIFFERENTIATION	**RATIONALIZATION**
up-scaling	generalization
autonomy	deconditioning
INDIVIDUALIZATION	**DOMESTICATION**
dependency	conditioning

Fig. 5.2 Paradoxes of modernization. *Source*: Own translation of van der Loo and van Reijen (1992, p. 40)

in-sourcing takes place among and between people and institutions but also increasingly between people and machines" (Koch 2012, p. 11). Domestication, finally, means simultaneously deconditioning from natural constraints like surmounting of physical distances due to technological artifacts (e.g. e-banking, e-learning) and conditioning towards these technologies (it is nowadays almost impossible to survive socially without a computer, the Internet, a cell phone, etc.).

There are many more empirical and theoretical examples which advocate a complementary coupling of individual and social scales as well as local and spatial scales. Scholz (2002), for example, proposes a theory of fragmented development, ranging from global to local fragmentation of social and economic activities. He argues that globalization implies simultaneously both homogenization and heterogenization which led to decreased relevance of the national scale but increased relevance of the global and local scale. A universal network of global cities is dominating the economies of regions almost everywhere. The term "global city", however, does not mean that the entire city is part of that global network, but only highly localized places like for example the Docklands and the traditional financial district in London.

Contemporary urban social geography also debates about issues which recognize trans-scaling phenomena. Knox and Pinch (2006) in their book address topics like "patterns of sociospatial differentiation", "spatial and institutional frameworks", "segregation and congregation", "neighborhood, community and the social construction of place", and "residential mobility and neighborhood change" as crucial for modern western societies and thus illustrate the significance of extending agent-based geosimulation techniques by the above mentioned macro-scales.

Apart from other methodological (e.g. resolution; see Walsh et al. 2004) and theoretical topics (e.g. complexity theory; see Easterling and Polsky 2004), it is the relative nature of scale which does not predetermine a specific level of domain, e.g. the individual and local in residential segregation models, and which emerges due to the conflation of different human-social relationships embedded in different local-spatial contexts. Swyngedouw (2004, p. 132f) points to these characteristics of scale from a political point of view: "Scalar configurations [on the one hand; A.K.] [...] are always already a result, an outcome of the perpetual movement of the flux of sociospatial and environmental dynamics. [...]. Spatial scales [on the other hand; A.K.] are never fixed, but are perpetually redefined, contested and restructured in terms of their extent, content, relative importance and interrelations. [...] These sociospatial processes change the importance and role of certain geographical scales, [...] and on occasion create entirely new scales".

All variation along the socio-spatial axis of scales is intrinsically tied to time. The simple successive nature of creating order with time becomes a complex concern as time is mutually interrelated with the social and spatial conceptualization of order. Table 5.1 illustrates the three dimensions in a simple linear fashion, but one can easily imagine that each line of the three domains is connected with all lines of the other two (the temporal structuration is according to Bossel 2007, the social according to Luhmann 1993, and the spatial according to McMaster and Sheppard 2004). It is obvious that no single model is able to incorporate all levels with

Table 5.1 Ranges of temporal, social and spatial scales

Temporal scale	Spatial scale	Social scale
Process (cause-effect)	Human body, household	Interaction system, organization
Feedback	Neighborhood	–
Adaptation, self-organization, evolution	City, metropolitan area, province/state, nation, continent, globe	Functionally differentiated systems (e.g. politics, economy, religion, science)

Source: Bossel (2007), McMaster and Sheppard (2004), Luhmann (1993)

all relations, but it would be worth extending pure bottom-up model approaches in geosimulation with top-down approaches, for example by coupling an agent-based sub-model with a system dynamics sub-model. With respect to archaeological models it makes a difference whether one is interested in a local, excavated village in order to reconstruct social structures and the date of the buildings, or if one is interested in the interregional transportation infrastructure of an expanding society over a long period.

5.4 ABM and GIS: Coupling Techniques of Different Methodological Domains

Agent-based modeling (ABM) and Geographic Information Systems (GIS) have both their strengths and their weaknesses when dealing with socio-spatial processes. While ABM intrinsically focuses on temporal processes of mobile (e.g. human) and immobile (parcels) agents but represents geographical space in a cursory manner (Benenson and Torrens 2004, p. 6), GIS offers extensive tools to represent and visualize space but is less suitable to integrate time as a continuous and inherent factor. There have been several attempts in the past decade or so to overcome these respective problems by either coupling ABM with GIS or by embedding one domain into the other. Crooks and Castle (2012) present a comprehensive and current overview of these strategies and concrete applications. This section therefore reviews briefly the most relevant methodological issues.

GI systems are able to store temporal information in several ways (Crooks and Castle 2012, p. 222): The location-based approach stores polygon or raster cell states for each time step, regardless of whether spatial change has occurred or not. Conversely, the time-based approach refers to a given time interval and here spatial change is related to a threshold of significance, i.e. change is only stored if it is sufficiently significant. The entity-based approach is directed towards the shape of spatial entities and their potential changes but not towards the attributes represented by these. Though the relevance of time is highly appreciated within GIS communities a true process representation of it has not yet been satisfactorily achieved (Reitsma and Albrecht 2006).

Agent-based modeling tools on the other hand have begun to implement some GIS functionality in varying degrees. Precursors of this development have been (and still are) Cellular Automata approaches which provide large modeling flexibility if space is conceptualized as a discrete raster grid and neighborhood rules play a dominant role in socio-spatial simulation (Iltanen 2012; Liu and Feng 2012). Software packages as, for example, NetLogo (Wilensky 1999) are able to import raster and vector data for visualization and analysis; polygon representations, however, are not handled as agents and agents' locational information is based on cell centroids rather than cartographic coordinates. Nevertheless, most (open source) ABM software is developing at a fast pace (Crooks and Castle 2012, p. 233ff) towards a more advanced insertion of GIS functionality.

Earlier stage attempts of linking GIS with ABM software have been state-of-the-art. While loose coupling "[...] usually involves the asynchronous operation of functions within each system, with data exchanged between systems in the form of files" (Crooks and Castle 2012, p. 224), close coupling "[...] is characterized by the simultaneous operation of systems allowing direct inter-system communication during the programme execution" (Crooks and Castle 2012, p. 225).

Meanwhile, sophisticated geosimulation systems are available, both within GIS and ABM. A central force in this respect has been object orientation in programming and in database storage. "The recent availability of an object-oriented approach to composing GIS software, which parallels the structure of ABM programming tools and uses common architectures like the Component Object Model (COM), has facilitated the integration of ABM and GIS tools" (Johnston 2012, p. 9). Crooks and Castle (2012, p. 233ff) provide an overview on ABM software systems like Swarm, MASON, Repast, NetLogo or AnyLogic, to mention just a few; their fields of application vary significantly and thus their capacities for integrating GIS functionality. They conclude: "[...] it needs to be noted that while such tools exist, integrating GIS data for ABM is still a difficult process [...] and many considerations are needed such as what data is needed, how should the data be utilised, how should agents interact with the data, etc." (Crooks and Castle 2012, p. 245).

For one of the most widely used GIS software packages, ArcGIS, an extension called Agent Analyst has been available for around a decade (it should be mentioned that while Agent Analyst is open source, ArcGIS is a commercial product of ESRI). "Agent Analyst provides an interface that can integrate the functionality of the Repast ABM software libraries with the geoprocessing environment in ArcGIS" (Johnston 2012, p. 11). Geoprocessing synchronizes agent states with mapping and cartographic visualization and allows the whole range of geospatial analysis. It is built as middleware between Repast ABM and ArcGIS. "Repast is used for the creation of the agent rules, object support, and scheduling. ArcGIS is used for data creation, GIS analysis, and display of the simulations" (Johnston 2012, p. 15). This means that generic agents (i.e., mobile human agents) and polygon agents (i.e., spatial agents) are treated as geo-referenced objects.

Thus, a plenitude of ever more advanced software systems are available providing data handling, simulation rules, geospatial analysis techniques, and high resolution visualization capabilities in one product.

5.5 Conclusion

Agent-based geosimulation modeling offers a particular kind of scientific reasoning and implies a specific style of thinking (Axelrod 1997; Resnick 1997). It enables a different access to heuristics, approaching the problem at hand in a way that goes beyond a textual description and explanation, a mathematical formulation, and a graphical visualization, though all these approaches are necessarily included. Geosimulation takes insights of complexity theory seriously into account (Manson et al. 2012) and recognizes interrelations between human and locational agents as crucial forces for our understanding of society and space. Furthermore and according to the "generativist" claim, it embeds socio-spatial relationships in different temporal contexts. Epistemologically, the work of translating from thoughts into formalized programming codes reveals a peculiar opportunity which has been presented by Hegselmann (2012, p. 6) as a continuous flux between scales: "It is a frequent programming experience that code [. . .] unintentionally (!) realises the particular as an instance of a generalisation that goes far beyond the original feature. [. . .] Almost never can we implement all alternatives. But from now on we know that whatever we implement is just an instance of something more general that might be called a migration regime".

Modeling of spatial processes, in an environment of geosimulating software tools, considers—in social science research domains—scales of different type and range. Alterations in space are always a result of, and a meaning for, social changes at different levels and within different temporal regimes. If this is true then the well-known geospatial analytic problem of the modifiable areal unit problem (MAUP; see Openshaw 1984) should be extended to the modifiable temporal unit problem (MTUP) and the modifiable social unit problem (MSUP), too (Koch and Carson 2012). There is no pre-given or pre-determined composition of social, spatial, and temporal scales which is to be used in a specific research problem but which is to be justified by the model purpose and verified and validated against empirical and theoretical knowledge.

Against this background, the use of geosimulation in archaeology provides numerous opportunities since the focus of this discipline is also on interdependencies between spatial, social, and cultural processes. Local knowledge (Geertz 1983) of ancient societies is imprinted in local geographies of sites and infrastructures helping geographers and archaeologists to understand socio-spatial conditions at that time. With respect to research in community resilience an appropriate link between geosimulation and archaeology is finally being suggested. In his books "Collapse" (2004) and "The World Until Yesterday" (2012) Jared Diamond comprehensively describes the techniques of socio-cultural and socio-spatial survival of traditional societies and tribes despite environmental perturbations. The scale link is present here as well. Community resilience can be defined as the "[...] existence, development, and engagement of community resources by community members to thrive in an environment characterized by change, uncertainty, unpredictability, and surprise" (Magis 2010, p. 402). The conflation of domains and scales becomes apparent: "Community resilience is the sum of neither resilience potentials of its members nor its environment. Through emergence, structures of resilience are generated at this level which in turn feed back to the local/micro units. This scale-dependent circularity also means that capabilities of crisis management cannot be simply generalised and transferred between scales. Rather, it is a more or less specific coupling of resources and constraints, of capabilities, skills, and resilience mechanisms which leads to more or less specific perceptions, assumptions, and proposals about how to deal with crises, vulnerabilities or risks" (Koch 2012, p. 16). A geosimulation model appears to be conceivable which integrates internal relations of human agents' communities dealing with external creeping transformations and/or sudden shocks by embedding these social systems into adequate (reliable) spatial settings.

Geosimulating archaeological facts or assumptions of socio-spatial structures and processes in varying social, spatial, and temporal scale seems to be a straightforward development in archaeological simulation, though, as Lake (2010, p. 17) states, it "is a minority activity and is likely to remain so for the foreseeable future". Aldenderfer (2010) takes a stand for the use of geosimulation in archaeology in order to suitably link the visualization domain with the knowledge domain. This link subsumes a couple of further reasons as, among others, an integration of large amounts of data of different nature (spatial, social, individual) and of different origin (empirical survey, remote-sensing), an advanced testing of hypotheses by use of visual and (geo-)statistical tools, and—maybe most importantly—the ability to explore dynamic processes. Geosimulation "is explicitly dynamic and, as such, provides a more realistic means by which to use space as not only a frame for action but one that has the capacity to directly modify agent behavior" (Aldenderfer 2010, p. 61).

Geosimulation, with its sophisticated analytical and visualization tools, might thus provide archaeological computational modeling efforts with opportunities to deal with methodological and theoretical problems that had been arisen in the past (Costopoulos et al. 2010; Costopoulos 2010; Lake 2010; Wobst 2010).

References

Aldenderfer M (2010) Seeing and Knowing: On the Convergence of Archaeological Simulation and Visualization. In: Costopoulos A, Lake MW (eds) Simulating Change: Archaeology into the Twenty-First Century. The University of Utah Press, Salt Lake City, pp 53–68

Axelrod R (1997) Advancing the Art of Simulation in the Social Sciences. In: Conte R, Hegselmann R, Terna P (eds) Simulating Social Phenomena. Lecture Notes in Economics and Mathematical Systems, vol 456. Springer, Berlin/Heidelberg/New York, pp 21–40

Batty M (2005) Cities and Complexities: Understanding Cities with Cellular Automata, Agent-Based Models, and Fractals. The MIT Press, Cambridge/London

Belting H (2005) Das Echte Bild: Bildfragen Als Glaubensfragen (The True Image: Issues of Images as Issues of Belief). C.H. Beck, Munich

Benenson I, Torrens PM (2004) Geosimulation: Automata-Based Modeling of Urban Phenomena. Wiley, New York

Berger JF, Nuninger L, van der Leeuw SE (2007) Modeling the role of resilience in socioenvironmental co-evolution: the Middle Rhône Valley between 1000 BC and AD 1000. In: Kohler TA, van der Leeuw SE (eds) The Model-Based Archaeology of Socionatural Systems. School for Advanced Research Press, Santa Fe, NM, pp 41–59

Bossel H (2007) Systems and Models: Complexity, Dynamics, Evolution, Sustainability. Books on Demand, Norderstedt

Bruch E (2006) Residential Mobility, Income Inequality, and Race/Ethnic Segregation in Los Angeles. In: Annual Meeting of the Population Association of America. http://paa2006. princeton.edu/papers/60143

Conte R, Hegselmann R, Terna P (eds) (1997) Simulating Social Phenomena. Lecture Notes in Economics and Mathematical Systems, vol 456. Springer, Berlin/Heidelberg/New York

Costopoulos A (2010) For a Theory of Archaeological Simulation. In: Costopoulos A, Lake MW (eds) Simulating Change: Archaeology into the Twenty-First Century. The University of Utah Press, Salt Lake City, pp 21–27

Costopoulos A, Lake MW (eds) (2010) Simulating Change: Archaeology into the Twenty-First Century. The University of Utah Press, Salt Lake City

Costopoulos A, Lake MW, Gupta N (2010) Introduction. In: Costopoulos A, Lake MW (eds) Simulating Change: Archaeology into the Twenty-First Century. The University of Utah Press, Salt Lake City, pp 1–8

Crooks AT (2008) Constructing and Implementing an Agent-Based Model of Residential Segregation Through Vector GIS. UCL Working Papers Series, Paper 133. Centre for Advanced Spatial Analysis, London. http://discovery.ucl.ac.uk/15185/

Crooks AT (2012) The Use of Agent-Based Modelling for Studying the Social and Physical Environment of Cities. In: de Roo G, Hiller J, van Wezemael J (eds) Complexity and Planning: Systems, Assemblages and Simulations. Ashgate, Burlington, pp 360–393

Crooks AT, Castle CJE (2012) The Integration of Agent-Based Modelling and Geographic Information for Geospatial Simulation. In: Heppenstall A, Crooks AT, See LM, Batty M (eds) Agent-Based Models of Geographical Systems. Springer, Dordrecht/Heidelberg/London/ New York, pp 219–251

Crooks AT, Heppenstall AJ (2012) Introduction to Agent-Based Modelling. In: Heppenstall A, Crooks AT, See LM, Batty M (eds) Agent-Based Models of Geographical Systems. Springer, Dordrecht/Heidelberg/London/New York, pp 85–105

Dean JS, Gumerman GJ, Epstein JM, Axtell RL, Swedlund AC, Parker MT, McCarroll S (2000) Understanding Anasazi Culture Change Through Agent-Based Modeling. In: Kohler TA, Gumerman GJ (eds) Dynamics in Human and Primate Societies: Agent-Based Modeling of Social and Spatial Processes. Oxford University Press, Oxford/New York

Diamond J (2004) Collapse: How Societies Choose to Fail or Succeed. Viking Adult, New York

Diamond J (2012) The World Until Yesterday: What Can We Learn from Traditional Societies. Allen Lane, London

Easterling WE, Polsky C (2004) Crossing the Divide: Linking Global and Local Scales in Human-Environment Systems. In: Sheppard E, McMaster RB (eds) Scale & Geographic Inquiry: Nature, Society, and Method. Blackwell Publishing, Malden/Oxford/Carlton, pp 66–85

Epstein JM (2006) Generative Social Science. Studies in Agent-Based Computational Modeling. Princeton University Press, Princeton/Oxford

Epstein JM, Axtell R (1996) Growing Artificial Societies: Social Science from the Bottom Up. Brookings Press/MIT Press, Washington/Cambridge/London

Fischer EP, Wiegandt K (2012) Dimensionen der Zeit: Die Entschleunigung unseres Lebens. Fischer Taschenbuch Verlag, Frankfurt am Main

Fossett M, Senft R (2004) SimSeg and Generative Models: A Typology of Model-Generated Segregation Patterns. In: Proceedings of the 2004 Conference on Social Dynamics: Interaction, Reflexivity, Emergence, pp 39–78

Fossett M, Waren W (2005) Overlooked implications of ethnic preferences for residential segregation in agent based models. Urban Stud 42(11):1893–1917

Geertz C (1983) Local Knowledge. Fontana Press, London

Giddens A (1990) The Consequences of Modernity. Stanford University Press, Stanford

Gigerenzer G (2007) Gut Feelings: The Intelligence of the Unconscious. Viking Adult, New York

Gimblett RH (ed) (2002) Integrating Geographic Information Systems and Agent-Based Modeling Techniques for Simulating Social and Ecological Processes. Oxford University Press, Oxford/New York

Gimblett RH, Skov-Petersen H (eds) (2008) Monitoring, Simulation, and Management of Visitor Landscapes. The University of Arizona Press, Tucson

Han J, Kamber M (2006) Data Mining. Concepts and Techniques, 2nd edn. Elsevier, San Francisco

Hegselmann R (2012) Thomas C. Schelling and the computer: some notes on Schelling's essay "on letting a computer help with the work". J Artif Soc Soc Simul 15(4):9. http://jasss.soc.surrey.ac.uk/15/4/9.html

Heppenstall A, Crooks AT, See LM, Batty M (2012) Agent-Based Models of Geographical Systems. Springer, Dordrecht/Heidelberg/London/New York

Holland JH (1998) Emergence: From Chaos to Order. Oxford University Press, Oxford

Ihde D (2006) Models, Models Everywhere. In: Lenhard J, Küppers G, Shinn T (eds) Simulation: Pragmatic Construction of Reality. Springer, Dordrecht, pp 79–86

Iltanen S (2012) Cellular Automata in Urban Spatial Modelling. In: Heppenstall A, Crooks AT, See LM, Batty M (eds) Agent-Based Models of Geographical Systems. Springer, Dordrecht/Heidelberg/London/New York, pp 69–84

Janssen MA (2010) Population aggregation in ancient arid environments. Ecol Soc 15(2). http://www.ecologyandsociety.org/vol15/iss2/art19/

Johnson S (2001) Emergence. Scribner, New York/London/Toronto/Sydney

Johnston K (2012) Agent Analyst: Agent-Based Modeling in ArcGIS. EsriPress, Redlands

Johnston K, Ver Hoef JM, Krivoruchko K, Lucas N (2001) Using ArcGIS Geostatistical Analyst. EsriPress, Redlands

Kennedy WG (2012) Modelling human behaviour in agent-based models. In: Heppenstall A, Crooks AT, See LM, Batty M (eds) Agent-Based Models of Geographical Systems. Springer, Dordrecht/Heidelberg/London/New York, pp 167–179

Knox P, Pinch S (2006) Urban Social Geography: An Introduction, 5th edn. Pearson, Harlow

Koch A (2009) Agents in space meet spatial agents: modeling segregation as a complex spatially explicit phenomenon. Eur J GIS Spat Anal 19(4):471–491

Koch A (2012) The Logics of Change: A Relational and Scale-Sensitive View on Poverty, Place, Identity, and Social Transformation. In: Kapferer E, Koch A, Sedmak C (eds) The Logics of Change. Poverty, Place, Identity, and Social Transformation Mechanisms. Cambridge Scholars Publishing, Newcastle upon Tyne

Koch A, Carson D (2012) Spatial, Social and Temporal Scaling in Sparsely Populated Areas - Geospatial Mapping and Simulation Techniques to Investigate Social Diversity. In: Jekel T, Car A, Strobl J, Griesebner G (eds) GI_Forum 2012: Geovisualization, Society and Learning. Wichmann, Heidelberg, pp 44–53

Kohler TA, Gumerman GJ (eds) (2000) Dynamics in Human and Primate Societies. Agent-Based Modeling of Social and Spatial Processes. Oxford University Press, Oxford/New York

Kohler TA, van der Leeuw SE (eds) (2007) The Model-Based Archaeology of Socionatural Systems. School for Advanced Research Resident Scholar Book, Santa Fe

Kohler TA, Kresl J, van West C, Carr E, Wilshusen RH (2000) Be There Then: A Modeling Approach to Settlement Determinants and Spatial Efficiency Among Late Ancestral Pueblo Populations of the Mesa Verde Region, U.S. Southwest. In: Dynamics in Human and Primate Societies: Agent-Based Modeling of Social and Spatial Processes. Oxford University Press, Oxford/New York

Küppers G, Lenhard J, Shinn T (2006) Computer Simulation: Introductory Essay. In: Lenhard J, Küppers G, Shinn T (eds) Simulation. Pragmatic Construction of Reality. Springer, Dordrecht, pp 1–22

Lake MW (2000) MAGICAL Computer Simulation of Mesolithic Foraging. In: Kohler TA, Gumerman GJ (eds) Dynamics in Human and Primate Societies: Agent-Based Modeling of Social and Spatial Processes. Oxford University Press, Oxford/New York, pp 107–143

Lake MW (2010) The Uncertain Future of Simulating the Past. In: Costopoulos A, Lake MW (eds) Simulating Change: Archaeology into the Twenty-First Century. The University of Utah Press, Salt Lake City, pp 12–20

Latour B (2005) Reassembling the Social: An Introduction to Actor-Network-Theory. Oxford University Press, Oxford

Laurie AJ, Jaggi NK (2003) Role of 'vision' in neighborhood racial segregation: a variant of the Schelling segregation model. Urban Stud 40(13):2687–2704

Leuangthon O, Khan DK, Deutsch CV (2008) Solved Problems in Geostatistics. Wiley, New York

Liu Y, Feng Y (2012) A Logistic Based Cellular Automata Model for Continuous Urban Growth Simulation: A Case Study of the Gold Coast City, Australia. In: Heppenstall A, Crooks AT, See LM, Batty M (eds) Agent-Based Models of Geographical Systems. Springer, Dordrecht/Heidelberg/London/New York, pp 643–662

Lloyd C (2007) Local Models for Spatial Analysis. CRC Press Taylor & Francis, Boca Raton/London/New York

Luhmann N (1993) Soziale Systeme: Grundriß einer allgemeinen Theorie. Suhrkamp, Frankfurt am Main

Magis K (2010) Community resilience: an indicator of social sustainability. Soc Nat Resour 23:401–416

Maguire DJ, Batty M, Goodchild MF (eds) (2005) GIS, Spatial Analysis, and Modeling. ESRI Press, Redlands

Manson SM, Sun S, Bonsal D (2012) Agent-Based Modeling and Complexity. In: Heppenstall A, Crooks AT, See LM, Batty M (eds) Agent-Based Models of Geographical Systems. Springer, Dordrecht/Heidelberg/London/New York, pp 125–139

McMaster RB, Sheppard E (2004) Scale and Geographic Inquiry. In: Sheppard E, McMaster RB (eds) Scale & Geographic Inquiry: Nature, Society, and Method. Blackwell Publishing, Malden/Oxford/Carlton, pp 1–22

Miller HJ, Han J (2009) Geographic Data Mining and Knowledge Discovery, 2nd edn. CRC Press, Boca Raton/London/New York

Mithen SJ (1994) Simulating Prehistoric Hunter-Gatherers. In: Gilbert N, Doran J (eds) Simulating Societies: The Computer Simulation of Social Phenomena. UCL Press, London, pp 165–193

Moss S, Davidsson P (eds) (2001) Multi-Agent-Based Simulation. Lecture Notes in Artificial Intelligence, vol 1979. Springer, Berlin/Heidelberg

Openshaw S (1984) The Modifiable Areal Unit Problem. GeoBooks, Norwich

O'Sullivan D, Millington J, Perry G, Wainwright J (2012) Agent-Based Models - Because They're Worth it? In: Heppenstall A, Crooks AT, See LM, Batty M (eds) Agent-Based Models of Geographical Systems. Springer, Dordrecht/Heidelberg/London/New York, pp 109–123

Pancs R, Vriend NJ (2007) Schelling's spatial proximity model of segregation revisited. J Public Econ 91:1–24

Patel A, Crooks AT, Koizumi N (2012) Slumulation: an agent-based modeling approach to slum formations. J Artif Soc Soc Simul 15(4). http://jasss.soc.surrey.ac.uk/15/4/2.html

Rao AS, Georgeff MP (1995) BDI-Agents: From Theory to Practice. In: Proceedings of the First International Conference on Multi-Agent-Systems. AAAI Press, Menlo, pp 312–319

Reitsma F, Albrecht J (2006) A Process-Oriented Data Model. In: Drummond J, Billen R, Forrest D, ao EJ (eds) Dynamic and Mobile GIS: Investigating Change in Space and Time. Taylor & Francis, London, pp 77–87

Remy N, Boucher A, Wu J (2009) Applied Geostatistics with SGeMS. Cambridge University Press, Cambridge

Resnick M (1997) Turtles, Termites, and Traffic Jams: Explorations in Massively Parallel Microworlds. MIT Press, Cambridge/London

Schelling TC (1969) Models of Segregation. In: American Economic Review, Papers and Proceedings, vol 59, pp 488–493

Schelling TC (1971) Dynamic models of segregation. J Math Sociol 1(2):143–186

Schmidt B (2002) Modelling of Human Behaviour: The PECS Reference Model. In: Proceedings of the 14th European Simulation Symposium, SCS Europe BVBA

Scholz F (2002) Die Theorie der fragmentierenden Entwicklung. Geographische Rundschau 54(10):6–11

Searle JR (2010) Making the Social World: The Structure of Human Civilization. Oxford University Press, Oxford

Sennett R (2012) Together: The Rituals, Pleasures and Politics of Cooperation. Yale University Press, New Haven/London

Sichman JS, Conte R, Gilbert N (eds) (1998) Multi-Agent Systems and Agent-Based Simulation. Lecture Notes in Artificial Intelligence, vol 1534. Springer, Berlin/Heidelberg

Smith EA, Choi JK (2007) The Emergence of Inequality in Small-Scale Societies: Simple Scenarios and Agent-Based Simulations. In: Kohler TA, van der Leeuw SE (eds) The Model-Based Archaeology of Socionatural Systems. School for Advanced Research Resident Scholar Book, Santa Fe, pp 105–119

Squazzoni F (2012) Agent-Based Computational Sociology. Wiley, New York

Stachowiak H (1973) Allgemeine Modelltheorie. Springer, Wien/New York

Swyngedouw E (2004) Scaled Geographies: Nature, Place, and the Politics of Scale. In: Scale & Geographic Inquiry: Nature, Society, and Method. Blackwell Publishing, Malden/Oxford/Carlton, pp 129–153

Tate NJ, Atkinson PM (eds) (2001) Modelling Scale in Geographical Information Science. Wiley, New York

Troitzsch KG, Mueller U, Gilbert GN, Doran JE (eds) (1996) Social Science Microsimulation. Springer, Dordrecht/Heidelberg/London/New York

van der Loo H, van Reijen W (1992) Modernisierung. Projekt und Paradox. dtv, Munich

Walsh SJ, Crews-Meyer KA, Crawford TW, Welsh WF (2004) Population and Environment Interactions: Spatial Considerations in Landscape Characterization and Modeling. In: Scale & Geographic Inquiry: Nature, Society, and Method. Blackwell Publishing, Malden/Oxford/Carlton, pp 41–65

Wang F (2006) Quantitative Methods and Applications in GIS. CRC Press Taylor & Francis, Boca Raton/London/New York

Wilensky U (1999) NetLogo. Center for Connected Learning and Computer-Based Modeling, Northwestern University, Evanston, IL. http://ccl.northwestern.edu/netlogo/

Wobst MH (2010) Discussant's Comments, Computer Simulation Symposium, Society for American Archaeology. In: Costopoulos A, Lake MW (eds) Simulating Change. Archaeology into the Twenty-First Century. The University of Utah Press, Salt Lake City, pp 9–11

Wolfram S (2002) A New Kind of Science. Wolfram Media, Champaign

Chapter 6
Large Simulations and Small Societies: High Performance Computing for Archaeological Simulations

Xavier Rubio-Campillo

6.1 Introduction

Computational simulation of societies has found in archaeologists one of the most promising fields of applications, as can be observed in the number of related publications (Doran et al. 1994; Doran 1999; Lake 2000; Diamond 2002; Kohler and van der Leeuw 2007; Costopoulos and Lake 2010; Lake 2013). The reason of this interest can be found in the link between the basic objectives of archaeology and simulation. Archaeology attempts to understand human behavior through the detection and analysis of spatio-temporal patterns related to the location and type of found structures and objects. Uncertainty is inherent to this process, as any research based on data from the archaeological record should cope with it, both in time and space (Crema et al. 2010). Within this context, computer simulation is the perfect virtual lab because it is capable of dealing with mathematically intractable problems such as the interaction of complex human behavior (Galán et al. 2009). They are suited to explore different hypotheses capable of explaining detected patterns, as well as to validate them, at least within the context of the simulation model. Agent-Based Models is one of the most widely spread simulation techniques; its basic concept of entities with individual decision-making processes interacting within a common environment is well suited to explore the type of questions faced by archaeologists.

From a broader perspective a common trend of Agent-Based Models simulating human behavior is a gradual increase in their complexity (Bonabeau 2002). These research projects often have at their disposal a vast amount of data that can be used to define the value of several parameters, as well as the behavior of the different agents.

X. Rubio-Campillo (✉)
Barcelona Supercomputing Center, Barcelona, Spain
e-mail: xavier.rubio@bsc.es

© Springer International Publishing Switzerland 2015
G. Wurzer et al. (eds.), *Agent-based Modeling and Simulation in Archaeology*,
Advances in Geographic Information Science, DOI 10.1007/978-3-319-00008-4_6

The results are large scale simulations executed in High-Performance Computing (HPC) infrastructures (i.e. supercomputers), created to explore realistic scenarios, develop policy analysis and manage emergency situations, to cite some applications (Leitao et al. 2013; Macal et al. 2008).

The emergence of these large scale simulations brings also a theoretical issue: the intractability of Agent-Based Models understood as computational problems. From this perspective the cost of executing an ABM in a computer follows a exponential curve. This is crucial if we want to understand the emergent behavior of an ABM given any set of initial conditions. Within this perspective the computational cost for the solution will increase exponentially to the number of parameters of the model. The consequence of this property is that, at a given point, we will not be able to solve an increasingly complex ABM in a reasonable amount of time, regardless the computer we use. We can have better computers, but the requirements will be orders of magnitude larger than our capabilities.

There is an ongoing debate in the community of archaeologists developing ABMs in relation to this increasing complexity. Some works try to follow the trend, developing detailed models designed to test hypotheses on realistic scenarios. Simultaneously a different group of modelers are focused on theory building, creating abstract simulations with simpler scenarios (for this division see Lake 2001, 2013). The discussion is centered on the best way to use the efforts of the community, taking into account the fact that realistic models cannot be fully defined on the basis of the fragmented data collected by the discipline. In this sense, the development of large scale complex models with weak assumptions could strongly bias the interpretation of the archaeological record, thus decreasing the scientific quality of research (Premo 2010). Paradoxically the development of abstract models without realistic backgrounds can keep away most potential audience from simulations, thus preventing most archaeologists from using this interesting tool.

Regarding HPC one could think that this technology cannot really offer anything to archaeologists, given that it seems focused on boosting large scale ABMs with huge scenarios and millions of agents. The models used in archaeological research never have these requirements, so where is the need for additional computational power? This chapter will show that the benefits of HPC are not constrained to execute large simulations. In particular we will explore:

- How can we speed up the execution of a computer simulation of a small-scale society?
- Is it possible to improve Agent-Based Models with an increase on computational power?
- What pitfalls and issues do the introduction of HPC create?

Not only is HPC important for accelerating the performance of the simulations, but we also need a boost on computing capabilities in order to answer some of the methodological questions arised during recent years. Some deficiencies of archaeological ABMs are related to the lack of these capabilities, specially regarding realistic hypothesis testing models. An immediate improvement caused by the use of additional computers is a better exploration of parameter spaces, but there are others that can be even more interesting for the discipline. The introduction of advanced decision-making processes and the development of multiscale simulations could also improve the scientific quality of computer simulations developed in archaeology.

Next section defines background concepts needed for the chapter. This is followed by a discussion about the properties of Agent-Based Models and the potential approaches to executing them using HPC resources. This section also provides an overview of existing software and applications usable for achieving this task. Fourth section deals with the methodological issues solved (and generated) by the introduction of HPC. Finally, different concluding remarks are provided to summarize the ideas of the chapter, as well as future lines of research.

6.2 Background

Some technical concepts need to be defined before continuing the discussion. A general overview of an HPC infrastructure can be seen in Fig. 6.1:

- The *Central Processing Unit* (CPU) is the atom of any computing system. It is responsible for calculating arithmetical and logistical operations. Most existing

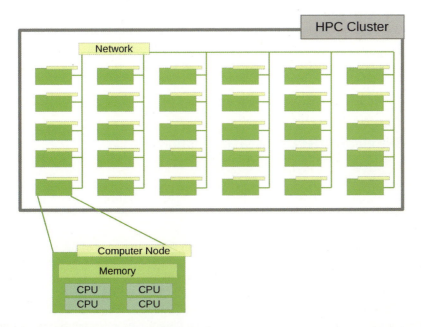

Fig. 6.1 General architecture of an HPC infrastructure

systems combine the power of different CPUs, and are known as multi-core processors.

- A *computer node* is a set of processors that share memory.
- An *HPC infrastructure* is a large set of computer nodes interconnected with some type of network. This can be as simple as some computers communicating through internet (i.e. grid computing), or a supercomputer where nodes are interconnected with high-speed networks.

Different hardware components can be used to execute a simulaton. If we want to distribute the process in more than one CPU we have two choices:

- Multiple processes in one node. The different CPUs of a computer node share the access to RAM (Random Access Memory). As a consequence, all of them can read and write in the same memory space. If a computer simulation wants to use more than one CPU it needs a mechanism to avoid conflicts generated by simultaneous writing of a particular memory section. This can be achieved with different techniques, being threads the most widely used. Another solution, OpenMP, is interesting for scientific applications given its capabilities for easily adding multicore execution to any software (Dagum and Menon 1998; Kuhn et al. 2000).
- Multiple nodes. The computer nodes of a distributed system (i.e. clusters, supercomputers, etc.) do not share memory. Any process being executed in more than one node must have some way to communicate among each section of the execution. This is achieved using the MPI (Message Passing Interface, MPI Forum 2009) protocol, that specifies how a distributed (or parallel) application must send and receive data using the network.

By definition, these solutions are complementary. State-of-the-art HPC applications use an hybrid approach to parallelism: a simulation is split into different nodes, while each section exploits all the cores of a given node.

6.3   Developing HPC-Based Simulations

From a technical point of view the distribution of an agent-based simulation in different computer nodes is a complex task. There is a strong, inherent coupling of the various components of the system. They are constantly interacting with each other, so communication is always intensive.

Each agent needs to gather knowledge from their environment, as well as from other agents in order to execute their decision-making processes. Once this phase is completed, there is a possibility that the agents will modify the environment (and so will the other agents). These mechanics translate, in terms of parallelization, to the need of sharing several layers of environmental data and agents between computer nodes, at every single time step. Furthermore, the execution of the agents' actions cannot usually be distributed within a single computer node (i.e. OpenMP), because there can be several conflicts related to agents accessing and modifying the same data at the same time.

Based on these constraints the best way to distribute the execution of an ABM over different computer nodes is to adopt a case-based approach. There is no optimal way to distribute an ABM, but a set of different solutions that can be applied to similar problems. The user should choose a particular method to split the execution amongst the nodes, depending on the nature of the model and the properties of the system. This idea is the basis of different initiatives like GridABM (Gulyás et al. 2011), a framework of template solutions for distributing these type of simulations.

Taking this perspective the majority of archaeological simulations share the following properties:

- Importance of environment. Hypothesis testing models developed in Archaeology usually focus on the strong relation between agents and the surrounding landscape. For this reason, these ABMs will need a substantial amount of spatially structured information, traditionally stored in a Geographical Information System (Conolly and Lake 2006). These platforms are spatial databases where data is stored as vectors (database records with spatial coordinates) and raster maps (bi-dimensional matrices containing regularly distributed values). The agents' environment will be modelled with a combination of these concepts, generating diverse information such as digital elevation models, site distribution, vegetation indexes, etc.
- Low number of agents. Past societies were not as dense as the ones we live in today. We will not need to reproduce situations with millions of agents, as simulations will usually model small-scale societies.
- Intensive local communication. A world without telecommunications means that, within a reasonable scale, an agent will only interact with the environment and agents that are spatially close to it. These interactions can be intense, but in any way an agent will need to know the entire situation of the world, just the section of it within a given interaction range.

6.3.1   Distribution of Archaeological ABM in Different Computer Nodes

These properties suggest that the best choice for splitting an archaeological ABM is to use spatial partitioning: each computer node owns a section of the entire simulated scenario, containing the different landscape data as well as the agents.[1] This is the solution used in Pandora (Rubio-Campillo 2013) and Repast-HPC (Collier and North 2011), the only HPC frameworks used in published archaeological ABMs.

---

[1] A different approach is seen in Long et al. (2011), where agents are grouped based on the network of interactions.

Fig. 6.2 Spatial partitioning
of an Agent-Based Model.
Each *color* represents the
section of the world owned by
a different computer node,
including raster maps with
environmental information
and agents. The zone where
different *colors* overlap is the
border zone

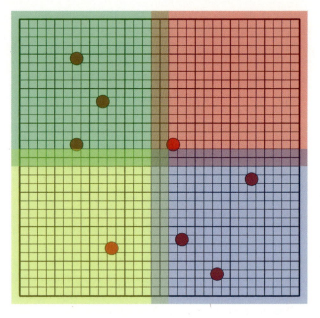

Pandora defines the environments where agents live as a set of layers containing raster map structures, following GIS standards. The world of the simulation is evenly divided among computer nodes, and each one of them will own a section of the environment, as well as the agents located in this part of the world. This layout is depicted in Fig. 6.2. Information in the border between adjacent nodes (raster maps and agents) will be communicated to neighbours every time step execution, in order to keep up-to-date data in the entire scenario. The size of this buffer border will be defined as the maximum interaction range of any agent, being the absolute horizon of actions of any component of the simulation. This approach is also adopted by Repast-HPC, and solves the issue of distribution for models of past societies without global interaction. The solution is highly scalable, given the fact that every computer node will need to communicate, at most, with eight neighbouring nodes (if nodes own rectangular regions), independently of the total size of the simulation.

Unfortunately, the issue of collision between agent's actions is still present, because two agents *living* in different computer nodes could be modifying the same bordering data at the same time. There are different techniques to avoid this conflict, but most of them can be computationally intensive (i.e. rollbacks). This overhead is affordable if the ABM is CPU intensive and the possibility of conflict is not high, but this is seldom the case.

Pandora, for example, takes a simpler approach. It is based in the segmentation of the spatial section owned by every computer node in four equal parts numbering 0–3, as it can be seen in Fig. 6.3. The agents contained in all 0 sections will be executed simultaneously without the possibility of conflicts, given the fact that they are not adjacent. Once all of them are finished, modified border data is sent to the

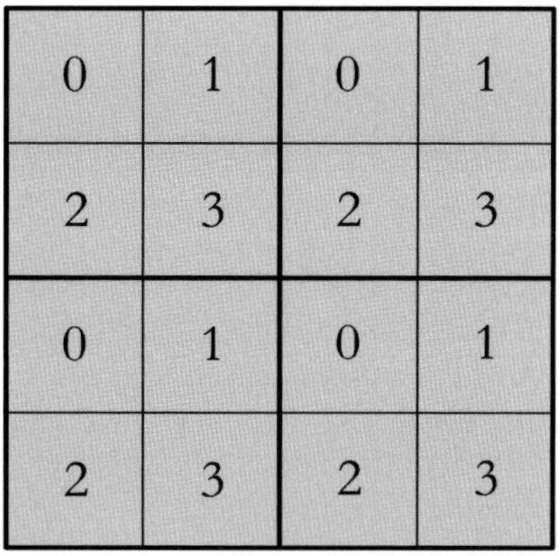

Fig. 6.3 The execution of any computer node is divided in four different sections, that are executed sequentially

0	1	0	1
2	3	2	3
0	1	0	1
2	3	2	3

neighbors, and a new section will begin its execution (1, 2 and finally 3). Once all of them are executed, the entire state of the simulation is serialized and a new time step can be evaluated.

The pitfall of this solution is that agents in section 0 will always be executed before agents in sections 1–3. Depending on the model the consequences of this effect can be nonexistent, or introduce artifacts in the outcome. As usual, a careful choice between the different strategies is needed, based on the existing scenario.

6.3.2   Simultaneous Execution of Agents

Parallelization of a simulation on different computer nodes is needed but is not sufficient. Every node should be able to use its complete set of CPU cores to simultaneously execute its agents. Again, the problem of conflicts between agents' actions is a barrier that must be broken if we want to avoid artifacts on simulation results.[2] To fix this we need to take a closer look at the way ABMs execute the set of agents (the scheduling system), in order to find properties useful to solve the problem.

Most of the time needed to execute an ABM is spent in the same task: the moment when the agents gather information, choose a particular set of behaviors, and execute them. These processes are always mixed in a single method, executed by every agent every time step. This is the approach taken by the three most popular ABM platforms: *tick* in Netlogo (Wilensky 1999), *step* in MASON (Luke 2011) and RePast (North et al. 2007).

It is worth to note that the first two processes do not modify anything, as the agent is only evaluating potential course of actions depending on existing data. For this

---

[2]For example two agents modifying at the same time the same cell of a given raster map.

reason we could simultaneously execute the decision-making process of different agents. This solution is safe of conflicts if the agents only choose a set of actions (without applying them).

Pandora uses this approach to split the step of an agent in three different methods. In the first one, *updateKnowledge*, an agent cannot modify the environment or other agents; it only gathers information. In the second one, *selectAction*, the agent executes the decision-making process and choose an action (it still cannot modify anything). Once every agent has chosen what it wants to do, Pandora executes the actions of the agents sequentially. Finally, the third method that a user can specify is *updateState*, where any agent can modify its internal state evaluating the results of its actions. This cycle *Explore–Decide–Apply* allows Pandora to distribute the execution amongst different CPU cores of a node, as the first two steps (the most computationally expensive) can be easily parallelized. The third one is executed sequentially, thus avoiding conflicts between the actions of the agents.

This structure could seem more complicated than just defining one method but, from a theoretical point of view, the division of an agent's execution in these three steps is more consistent than the traditional ABM approach. The single method implementation mixes the different stages of an agent's cycle, that should be correctly specified while building the model (see Fig. 6.4). Dividing the execution as shown here avoids this problem, forcing coherence during the transition from theory to code.

In addition this solution provides a cleaner interface to implement agents with advanced artificial intelligence, when this requirement is needed. The AI algorithm will be executed in one phase of the cycle (*decide*), that encapsulates the entire decision-making process and keeps it separated from the rest of the model.

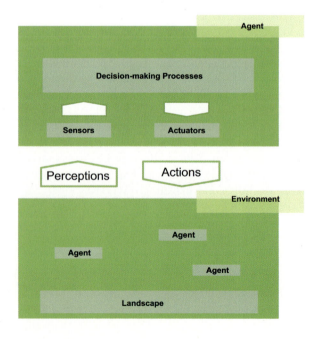

Fig. 6.4 The execution cycle of an agent

6.3.3 Serializing Distributed Simulations

If every CPU of every computer node is simultaneously executing agents there is a additional issue that needs to be dealt: bottlenecks during the serialization of the results. The scientific analysis of an ABM simulation will usually need to keep track of the evolution of the entire system from the beginning until the end of an execution. This means that the simulation needs to provide access to the state of the simulation at any given time step and at multiple levels of scale (from summary statistics to the state of individual agents). Besides, most HPC systems use a job cue where programs are submitted in order to be executed. This batch system is not interactive, so the researcher will usually access the outcome after the simulation finishes. The entire simulation state should be stored in a single file while it is running, in order to analyze it after the execution. The different nodes should wait to store their information while the rest are writing. This heavily affects the performance of the simulation, wasting resources and time due to the high volume of data to serialize.

One solution to the serialization bottleneck is that every computer node writes into a different file. The problem of this approach is that it requires postprocessing of different files in order to merge results, and this is also a costly operation. The optimal strategy, chosen by both frameworks (Repast-HPC and Pandora) is the distribution of the serialization through HDF5 (Folk et al. 1999). It is an open-source library specifically designed to store the outcome of scientific visualizations. It allows for simultaneously serializing data from several computer nodes in a structured, binary format. In addition, raster maps stored using HDF5 can be loaded by GIS applications (e.g. Quantum GIS, Sutton et al. 2009) in order to minimize the postprocessing of the data.

6.3.4 Theoretical Performance of Distributed ABMs

We can compute the improvements of distributing ABMs taking a look at theoretical wall time (real-world time needed to finish an execution). We can define the time needed for a model with local interaction as:

$$WallTime_L = N \cdot timeSteps \cdot (E + I \cdot L)/numCpus$$

being N the number of agents in the simulation, E the time an agent spends interacting with the environment in a given time step, I the time spent interacting with another agent and L the number of agents within range of interaction.[3] Similarly, the wall time for a model with global interaction is defined as:

$$WallTime_G = N \cdot timeSteps \cdot (E + I \cdot N)/numCpus$$

---

[3]This would be the cost for the majority of archaeological ABMs.

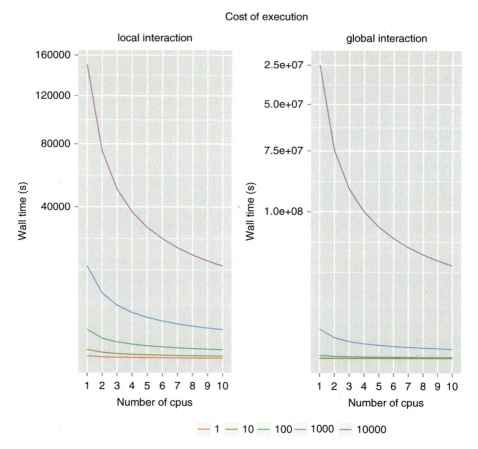

Fig. 6.5 Theoretical comparison of temporal costs of execution for Agent-Based Models with parameters $timeSteps = 1,000$, $E = 0.005$ s, $I = 0.001$ s, and $L = 10$ for (*left*) simulations where only local interaction exist and (*right*) simulations where agents interact with any other existing agent. Y axis shows (with a root squared scaling) the time needed to execute simulations with different number of agents (defined by *color*) for given cpus (X axis)

Figure 6.5 compares the theoretical cost of executing an ABM for both local and global interaction scenarios.

This decrease in wall time is the maximum boost in performance that a distributed simulation can achieve. Overheads generated by communication, serialization and other processes will lower performance, to the point that the increase in the number of computer nodes could slow the execution of very simple models. In this context the modeller is responsible for choosing the optimal infrastructure capable of accelerating the simulation execution (Wittek and Rubio-Campillo 2012b).

6.4 Computational Solutions of Methodological Problems

We will explore three methodological discussions that can be enriched with the introduction of increasing computing power: parameter sweeps, exploration of size dependent emergence, and the problem of self-fulfilling prophecies.

6.4.1 Parameter Sweeps

The first benefit from having a large set of CPUs is that you can execute a higher number of runs for a given experiment. Archaeological problems deal with a high degree of uncertainty, so a large percentage of models embody stochastic processes. The consequence is that we cannot analyze a simulation based on one run, as different executions provide different results; we need to execute several runs, and their number will increase depending on the degree of stochasticity of the model.

An HPC infrastructure minimizes the time needed to achieve this task but it is important to note that this problem has not been thoroughly explored, because the proper number of simulations is difficult to know; how many runs of a given scenario must be executed before being sure that the results are fully understood? A common technique is to compare the average and standard deviation of particular outcomes for different sets of the scenario (e.g. 10, 100 and 1,000 runs), but it still remains a tricky issue for most models. In any case this is a field that should be addressed in the near future.

If we want to grasp how the model is affected by different initial conditions we also need to execute parameter sweeps. The technique consists on exploring the combination of possible initial values of the simulation. This analysis allows the researcher to fully understand the importance of the parameters as well as the relation between them. If the number of parameters is large this could be an impossible task, as a correct analysis should explore the whole range of combinations; as a consequence the scientific quality of the model will decrease, because the effect of the different parameters in the outcome will not be fully understood.

This problem is even more important in archaeological models, as it is difficult to give realistic values for most parameters (given the mentioned uncertainty). Some models avoid this problem fitting real data to simulated data. The problem is that this is not always possible, because the solution requires high-quality archaeological data that usually is unavailable. As a result archaeology-related ABMs often suffer from the problem of having too many parameters that are nor explored neither calibrated.

One of the possible solutions is the adoption of HPC. This would allow, at a technical level, to execute the needed number of simulations in order to perform sensitivity analysis using different techniques (i.e. genetic algorithms, see Stonedahl and Wilensky 2010). Nevertheless the difficulties to understand the outcome remain, as the analysis of the results will be extraordinarily complex, and every new parameters will increase the problem at an exponential rate. Besides, HPCs are not designed to execute a large number of simple simulations. The cost of the interconnectivity between computer nodes is extremely high, and is not really used if every run is executed in a different node. In the end, a simple cluster or cloud-based services would have a better trade-off between cost and capabilities, being HPC best suited to face the following two issues.

6.4.2 Size Dependent Emergence

A quick glance at ABMs published in archaeological journals suggest that they tend to be small (for a survey of recent simulations see Lake 2013). The number of agents will seldom exceed the order of thousands, so the idea of having a large-scale simulation with zillions of agents do not seem related to the usual case studies. If a large process needs to be modeled it is easier to jump from one scale to a higher rank, thus adding a level of abstraction to the behavior of the agents. In the end the best practice is the choice of a resolution as lowest as possible; it will avoid the computational requirements and complications of huge simulations.

The pitfall of this approach is that some times the optimal spatial, temporal and behavioral resolutions do not fit well. Imagine the simulation of migration processes at a continental scale such as hominin dispersion from Africa. Models designed for this task have extreme low resolutions (hundreds of year per time step, thousands of kilometers per discrete spatial position). Given the temporal and spatial span it seems the best choice, and it has been adopted by several research works during the last decades (Mithen and Reed 2002; Nikitas and Nikita 2005; Hughes et al. 2007)

A closer look to the processes being modeled makes clear that the different resolutions do not fit: there is a gap between the choice of spatiotemporal scale and the scale of the behavior defined in the agents. On the one hand, space and time resolutions are chosen to model a process that is being developed at a huge scale. On the other hand, the behavior seems to be modeling a small-scale event (in the case of dispersion, the movement and reproduction of human groups). The truth is that behavior is not suited to the scale, because in reality during a time step any human would be able to move wherever he/she wants around the simulated world.

For example, in the cellular automata defined in the model Stepping Out (Mithen and Reed 2002, p. 436) a time step is defined as 250 years. During this time a cell can colonize its neighbors, being all of them triangles with sides of 322 km. Why a further cell cannot be accessed during this huge time interval? A cell at thousands of kilometers from the point of origin could be colonized; if this was not the case was for other reasons, not because it was physically impossible: this constraint is introduced in the model to deal with the divergence of the scales. In the end, a paradox emerges from the model: to understand the reasons why humans colonized entire continents we need to explore behavior at a small scale. If we want to correctly address the relation between behavior, spatial scale and time scale, we need to let the migration process emerge, instead of applying the behavior suitable for one scale to worlds defined at other scales while forcing artificial constraints.

A second issue related to the size of simulations arise from the concept of emergence, typical of ABMs: how can we model mechanisms that can only be understood for a large number of agents? For example, warfare tactics are strongly related to the size of an army. The tactics used in a given period are linked to the number of available soldiers that were deployed in a battlefield. Studies on battlefield archaeology of the eighteenth century (Rubio-Campillo et al. 2012) shows how linear warfare of this era cannot be correctly modeled while trying to simulate a few individuals. The reason is that the studied cultural traits (firing systems) were designed to be used with thousands of soldiers, given the lack of accuracy of existing fire weapons. Some of the hypotheses regarding command and control, and even individual behavior follow the same reasoning, as do other types of traits. In this context, we cannot simplify the simulation using fewer agents, as the behavior we want to explore is linked to the number of them that are interacting at a given time step.

Both issues (incorrect relation between scales and large-scale behaviors) can be included inside the general concept of size-dependent emergence. True enough, lots of emergent behaviors are detected in small worlds, but certain phenomena can only be observed in large-scale scenarios (Murphy 2011). HPC is essential in these cases, as we will need to create larger simulations, more costly in terms of computer power and more difficult to analyze if an HPC is not available.

6.4.3   Solving the Dilemma of Self-fulfilling Prophecies

The two previous topics proof that some models cannot be simplified or split in simpler simulations. As a consequence, the results can be more complex to understand and justify. Parallel to these issues there is another one that, in our opinion, seems to be the most important issue of state of the art archaeological simulations: the definition of behaviors.

The vast majority of published ABMs are based on the classic SugarScape model by Epstein and Axtell (1996). The model consists of a discrete world, defined as a finite bi-dimensional matrix, where a set of agents interact between them as well as with the environment. This behavior is defined as a list of rules; if one condition applies, a simple behavior is executed (e.g. if there is no food the agent will move to the adjacent place with more resources). In other words, the decision-making process of the agent is wired, as its choices are completely predefined. As a consequence it is incapable of finding solutions and reacting to conditions not devised by the modeler. This is a weakness of the methodology if we think that the interest on ABMs is precisely the emergence of large scale behaviors from the interaction of these simple rules. In theory these behaviors should not be explicitly defined in the original model, but as we said behavior is predefined, so there exist an important thread of circular explanations (Macal and North 2010).

Moreover, if everything in an ABM is related to the agent's behavior, why is it not analyzed like a parameter? Behavior is arbitrarily defined based on archaeological assumptions, so it should also be explored to ensure scientific quality. Most models do not face this issue, so we are not really sure if different behaviors (even slight variations) affect the observed outcome as no sensitivity analysis is performed to study it.

The consequence of this problem is that critiques against ABM focus on the fact that they are self-fulfilling prophecies: emergent traits are not related to the problem, but to the way it was programmed, and if we can take a look at the code, we will learn the implicit or explicit assumptions the modeler introduced in order to achieve the final outcome.

The solution of this issue is critical for the future of ABM. For simple models it can be argued that the assumptions are less important, because such models can be replicated and understood without problems. Theory building models guarantee the scientific quality, as the number of parameters is small and behavior is so simple that the emergence of non-expected behavior can be understood with proper analysis.

The same cannot be said about hypothesis testing simulations. They are usually more complex because their goal is to understand realistic scenarios. For this reason these models create agents with several traits and a large list of conditions and rules, as simpler agents would not be able to take decisions based on the amount of data being used. In the end it will become impossible to understand, even by the creator of the model, which system properties emerged from the simulations, and which ones from the way the agents were programmed. Even though the problem is often not explicited, its consequences are so important that we have to wonder if hypothesis testing models are really useful to understand social change, except for some approaches that use excellent datasets and simple mechanisms, like evolutionary archaeology (Premo 2010; Lake 2013).

Any solution to this issue needs to avoid the design of the traditional rule-based agents. Luckily enough Artificial Intelligence has developed, in the past decades, several alternatives to the modeling of decision-making processes. The change would provide access to a plethora of well-known methods that could be integrated in the agents' design such as goal oriented agents based on atomic actions.

Imagine that we want to explore the foraging strategies of Hunter-Gatherers in a realistic landscape. Instead of defining every way in which an agent can interact with the environment, we could define its basic goal as survival (i.e. getting each time step a given level of food) and a set of simple actions that it can use to achieve the goal. The only possible actions would be moving to another location or forage the place where the agent currently is. Every action would have potential associated rewards and costs, and the decision-making process would consist on the choice of a single action each time step. The whole process can be seen as a fully observable stochastic state model, or Markov Decision Process (MDP). In this context, the agent needs to explore possible options and the impact on its future, thus choosing the best one given its knowledge of the environment. Figure 6.6 shows this process, where the agent explores its future applying different sets of actions.

The benefit of this approach is that MDPs are well known models of decision-making in the field of Artificial Intelligence. There are different ways to solve it, like the family of A* algorithms or the UCT approach (Bonet and Geffner 2012).

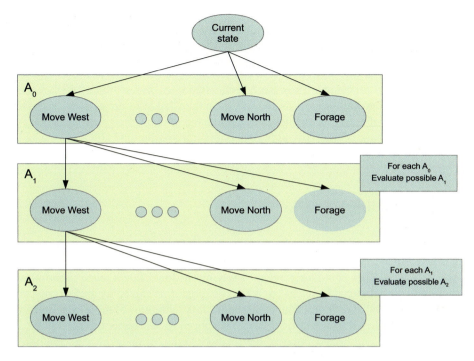

Fig. 6.6 Exploration of actions using a Markov Decision Process

This is a CPU consuming algorithm that acts as a planner for the agent: given a set of goals, the agent will simulate on its own the accumulated effects of executing different available actions. Thus, at every time step potential rewards and costs will be explored for a large set of possible actions, given a defined policy and a search depth. When the process is finished the agent will choose to execute the best action for the next state (when the algorithm will be executed again).

This change in the architecture of the agents solves the problem of implementing realistic scenarios at several levels:

1. Definition of the model. It is simpler to think on a small set of possible actions than a large set of rules. Moreover, if archaeological data is scarce these rules can not be defined at all (but we can still define the small actions).
2. Complex behavior without complex rules. Instead of defining rules the modeler defines goals and atomic actions. They are more understandable and for this reason the model can be best understood by archaeologists that did not implement the model.
3. Verification. The researcher has different available algorithms to solve the MDP. They are already understood and published by AI experts, so the modeler can use third-party implementations instead of developing new code.

4. Extension. The division of behavior between potential actions and decision-making process allows for the exploration of different approaches (e.g. what if we introduce a cognitive model, social interaction or partial information?).

To summarize this approach, in order to model realistic scenarios we need realistic decision-making processes. They have been researched by Artificial Intelligence during the last decades, so we should use their knowledge to create new types of agents, capable of acting in a realistic way while maintaining scientific quality. It is clear that this approach requires additional computing power, and HPCs are the key that will allow its introduction into archaeological ABMs.

6.5   Concluding Remarks

The use of simulation in science always require further computational power. Only high-performance computing (HPC) resources are capable of dealing with large simulation scenarios containing agents applying artificial intelligence algorithms with high computational costs. Even though the problems solved by archaeological ABMs are somewhat different than other fields, the capabilities of these systems can also be exploited. We have explored different options to accelerate the execution of the simulations, but also to improve the methodological framework and scientific quality of these models.

As we already know hardware infrastructures evolve at a fast pace; any software solution must be independent of the characteristics of the underlying system. On one side of the spectrum we find the supercomputers, where hundreds or thousands of nodes are located inside a facility with high-speed interconnection, forming a homogeneous and reliable environment. The increasing trend of cloud computing is present on the other side, where a theoretical unit of computing can, in reality, shape multiple and varied hardware infrastructures linked with medium or low-speed interconnections (Armbrust et al. 2009). The debate about which system is better is far from the aim of this chapter, but it is important to note that, from our perspective, they serve different purposes. Cloud computers are definitely needed when no supercomputing infrastructure is available. Even if a supercomputer is an option, cloud computers are more cheaper when the agents inside our model do not interact constantly. If the model is communication-intensive, on the other hand, a supercomputer is the best available choice, as its performance is optimal for this kind of executions (Wittek and Rubio-Campillo 2012a). The election of the platform used to execute an ABM will be increasingly difficult given the varied options, and the modellers should be able to choose based on the properties of their own simulations.

Additionally the constant improvement of available hardware components provides new ways for scientific simulation to exploit new computing capabilities. For example, Graphics Processor Units (GPUs) are increasingly being used to boost the performance of particular sections of a simulation (D'Souza et al. 2007). Even though it is unclear how small-scale ABMs could benefit from this approach, some examples have been suggested (Wittek and Rubio-Campillo 2013).

To conclude, this chapter provides an overview of the impact of HPC in archaeological ABMs. It is a decisive advance on the quality of the models focused on small-scale societies with a high demand for complex behaviors. The boost in computing capabilities provides solutions to different methodological issues derived from the use of this technique. In any case, we should be aware that, even if they are extremely powerful, they are just a tool; it is the responsibility of the researchers to develop models capable of exploiting these resources and solve interesting problems.

Acknowledgements Special thanks to Jose María Cela, Miguel Ramírez and two anonymous reviewers for their suggestions and comments on the topic and the preliminary versions of the text. The author is part of the SimulPast Project (CSD2010-00034) funded by the CONSOLIDER-INGENIO2010 program of the Spanish Ministry of Science and Innovation.

References

Armbrust M, Fox A, Griffith R, Joseph A, Katz R, Konwinski A, Lee G, Patterson D, Rabkin A, Stoica I (2009) Above the Clouds: A Berkeley View of Cloud Computing. Tech. Rep. UCB/EECS-2009-28, University of California, Berkeley

Bonabeau E (2002) Agent-based modeling: methods and techniques for simulating human systems. Proc Natl Acad Sci USA 99(3):7280–7287

Bonet B, Geffner H (2012) Action Selection for MDPs: Anytime AO* vs. UCT. In: Proceedings of the 26th AAAI Conference on Artificial Intelligence, pp 1749–1755

Collier N, North M (2011) Repast HPC: A Platform for Large-Scale Agent-Based Modeling. In: Dubitzky W, Kurowski K, Schott B (eds) Large-Scale Computing Techniques for Complex System Simulations. Wiley, Hoboken, pp 81–110

Conolly J, Lake M (2006) Geographical Information Systems in Archaeology. Cambridge University Press, Cambridge

Costopoulos A, Lake MW (eds) (2010) Simulating Change: Archaeology into the Twenty-First Century. The University of Utah Press, Salt Lake City

Crema E, Bevan A, Lake M (2010) A probabilistic framework for assessing spatio-temporal point patterns in the archaeological record. J Archaeol Sci 37(5):1118–1130

Dagum L, Menon R (1998) OpenMP: an industry standard API for shared-memory programming. Comput Sci Eng 5(1):46–55

Diamond J (2002) Life with the artificial Anasazi. Nature 419(6907):567–569

Doran J (1999) Prospects for agent-based modelling in archaeology. Archeologia e Calcolatori 10:33–44

Doran JE, Palmer M, Gilbert N, Mellars P (1994) The EOS Project: Modelling Upper Palaeolithic Social Change. In: Gilbert N, Doran J (eds) Simulating Societies. UCL Press, London, pp 195–221

D'Souza RM, Lysenko M, Rahman K (2007) Sugarscape on Steroids: Simulating Over a Million Agents at Interactive Rates. In: Proceedings of the Agent2007 Conference

Epstein JM, Axtell R (1996) Growing Artificial Societies: Social Science from the Bottom Up. Brookings Press/MIT Press, Washington/Cambridge/London

Folk M, Cheng A, Yates K (1999) HDF5: A File Format and i/o Library for High Performance Computing Applications. In: Proceedings of the 12th Conference on Supercomputing, Portland

Galán JM, Izquierdo LR, Izquierdo SS, Santos J, del Olmo R, López-Paredes A, Edmonds B (2009) Errors and artefacts in agent-based modelling. J Artif Soc Soc Simul 12(1):1. http://jasss.soc.surrey.ac.uk/12/1/1.html

Gulyás L, Szabó A, Legéndi R, Máhr T, Bocsi R, Kampis G (2011) Tools for Large Scale (Distributed) Agent-Based Computational Experiments. In: Proceedings of the Computational Social Science Society of America Annual Conference 2011

Hughes J, Haywood A, Mithen S, Sellwood B, Valdes P (2007) Investigating early hominin dispersal patterns: developing a framework for climate data integration. J Hum Evol 53: 465–474

Kohler TA, van der Leeuw SE (eds) (2007) The Model-Based Archaeology of Socionatural Systems. School for Advanced Research Press, Santa Fe

Kuhn B, Petersen P, O'Toole E (2000) OpenMP versus threading in C/C++. Concurrency Pract Exp 12(12):1165–1176

Lake MW (2000) MAGICAL Computer Simulation of Mesolithic Foraging. In: Kohler TA, Gumerman GJ (eds) Dynamics in Human and Primate Societies: Agent-Based Modelling of Social and Spatial Processes. Oxford University Press, New York, pp 107–143

Lake M (2001) Numerical Modelling in Archaeology. In: Brothwell D, Pollard A (eds) Handbook of Archaeological Science. Wiley, Chichester, pp 723–733

Lake M (2013) Trends in archaeological simulation. J Archaeol Method Theory. doi:10.1007/s10816-013-9188-1

Leitao P, Inden U, Rückemann CP (2013) Parallelising Multi-Agent Systems for High Performance Computing. In: INFOCOMP 2013: The Third International Conference on Advanced Communications and Computation, Portugal, pp 1–6

Long Q, Lin J, Sun Z (2011) Agent scheduling model for adaptive dynamic load balancing in agent-based distributed simulations. Simul Model Pract Theory 19(4):1021–1034

Luke S (2011) Multiagent Simulation and the MASON Library. George Mason University. http://cs.gmu.edu/~eclab/projects/mason/manual.pdf

Macal C, North M (2010) Tutorial on agent-based modelling and simulation. J Simul 4(3):151–162

Macal C, North M, Pieper G, Drugan C (2008) Agent-based modeling and simulation for exascale computing. SciDAC Rev 8:34–41. http://www.scidacreview.org/0802/index.html

Mithen S, Reed M (2002) Stepping out: a computer simulation of hominid dispersal from Africa. J Hum Evol 43(4):433–462

MPI Forum (2009) Message Passing Interface (MPI) Forum Home Page. http://www.mpi-forum.org/

Murphy J (2011) Computational Social Science and High Performance Computing: A Case Study of a Simple Model at Large Scales. In: Proceedings of the 2011 Annual Conference of the Computational Social Science Society of America

Nikitas P, Nikita E (2005) A study of hominin dispersal out of Africa using computer simulations. J Hum Evol 49:602–617

North M, Howe T, Collier N, Vos J (2007) A Declarative Model Assembly Infrastructure for Verification and Validation. In: Takahashi S, Sallach D, Rouchier J (eds) Advancing Social Simulation: The First World Congress

Premo LS (2010) Equifinality and Explanation: The Role of Agent-Based Modeling in Postpositivist Archaeology. In: Costopoulos A, Lake M (eds) Simulating Change: Archaeology into the Twenty-First Century. University of Utah Press, Salt Lake City, pp 28–37

Rubio-Campillo X (2013) Pandora: An hpc Agent-Based Modelling Framework. Software. https://github.com/xrubio/pandora

Rubio-Campillo X, María Cela J, Hernàndez Cardona F (2012) Simulating archaeologists? Using agent-based modelling to improve battlefield excavations. J Archaeol Sci 39:347–356

Stonedahl F, Wilensky U (2010) Evolutionary Robustness Checking in the Artificial Anasazi Model. In: Proceedings of the 2010 AAAI Fall Symposium on Complex Adaptive Systems

Sutton T, Dassau O, Sutton M, Nsibande L, Mthombeni S (2009) Quantum GIS Geographic Information System. Open Source Geospatial Foundation, Quantum GIS Development Team

Wilensky U (1999) NetLogo. Center for Connected Learning and Computer-Based Modeling, Northwestern University, Evanston, IL. http://ccl.northwestern.edu/netlogo/

Wittek P, Rubio-Campillo X (2012a) Military Reconstructive Simulation in the Cloud to Aid Bat-
    tlefield Excavations. In: Proceedings of the 4th International Conference on Cloud Computing
    Technology and Science, pp 869–874
Wittek P, Rubio-Campillo X (2012b) Scalable Agent-Based Modelling with Cloud HPC Resources
    for Social Simulations. In: Proceedings of the 4th International Conference on Cloud Comput-
    ing Technology and Science, pp 355–362
Wittek P, Rubio-Campillo X (2013) Social Simulations Accelerated: Large-Scale Agent-Based
    Modeling on a GPU Cluster. In: GPU Technology Conference, San Diego. Poster DD06. http://
    on-demand.gputechconf.com/gtc/2013/poster/pdf/P0197_PeterWittek.pdf

Part III
Applications

Chapter 7
Mining with Agents: Modelling Prehistoric Mining and Prehistoric Economy

Kerstin Kowarik, Hans Reschreiter, and Gabriel Wurzer

7.1   Introduction

Mining areas are characterized as centres of *production* and *consumption*. Aspects such as expert knowledge, intra- and superregional communication, the operation and maintenance of traffic and trade networks further add to its complexity and require consideration. All these interdependent conditions demand an analytic approach combining different levels of observation, both spatially and in context of the model used.

With its spatial modelling approach, Agent-Based Simulation (ABS) can aid a clear formalization with respect to one of these levels (within a mining gallery, considering the whole mine as a system, considering mining facility with its supporting settlements, considering a whole region, see Fig. 7.1). The spatial data basis that is used in this respect largely depends upon context (refer to Fig. 7.2):

- Continuous representations involve vector data coming from GIS or CAD, while discretized representations import per-cell raster data from GIS or bitmap images. To be precise, space in an ABS is a locality-based data source for a set of defined properties (or layers, in GIS jargon). An evaluation at spot (x,y) yields the values of properties $p_1 \ldots p_n$, where n is the number of layers present.

K. Kowarik (✉) • H. Reschreiter
Prehistory, Natural History Museum Vienna, Vienna, Austria
e-mail: kerstin.kowarik@nhm-wien.ac.at; hans.reschreiter@nhm-wien.ac.at

G. Wurzer
Vienna University of Technology, Wien, Austria
e-mail: gabriel.wurzer@tuwien.ac.at

© Springer International Publishing Switzerland 2015
G. Wurzer et al. (eds.), *Agent-based Modeling and Simulation in Archaeology*,
Advances in Geographic Information Science, DOI 10.1007/978-3-319-00008-4_7

Fig. 7.1 A multi-level simulation approach involves several stages of increasing complexity. ABS can be used to formalize models such that each level acts as a black box for the next-higher stages, thereby making results reusable and composable

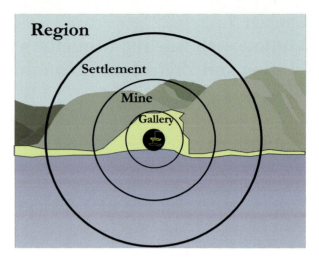

Fig. 7.2 ABS excels at spatial simulation, where space can be either continuous or discrete (i.e. consisting of cells). Space acts as a data source, providing properties $p_1 \ldots p_n$ at a certain location (x,y). Agents are movable entities within this environment that can query and alter these spatial properties. They also have an identity i and carry properties $q_1 \ldots q_m$ with them. Thus, different agents can be queried from the outside for their locality (and properties residing there) as well as their own properties, which makes complex interactions between agents and space possible

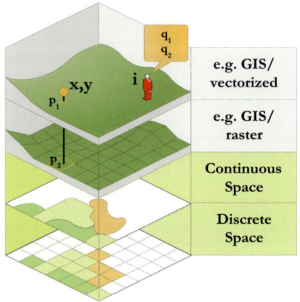

- Quite similarly, agents are movable entities residing at a certain location x,y. They hold their own properties $q_1 \ldots q_m$, which are typically defined per agent type (e.g. persons). Agents furthermore have an identity i that can be used to query a specific agent.

Even though literature sometimes speaks of "agents possessing a brain of their own", we see this as a completely separate matter: For us, agents are essentially moveable containers for data, governed by the actual simulation code that also influences all other parts of the environment such as the underlying space. The statement of the simulation model is therefore the choice "what to do" with all of these data-holding entities, which is exactly what we focus on with this contribution, giving prehistoric mining as an example case.

7.2 The Prehistoric Salt Mine of Hallstatt/Upper Austria

The prehistoric salt mines of Hallstatt are located in southern part of Upper Austria in the alpine Dachstein region. The mining areas as well as the famous Early Iron age cemetery lie 400 m above the historical mining town of Hallstatt, in the Salzberg Valley. The topographic and geographic situation can be described as difficult to access and remote. The contemporaneous settlement areas were located about 30 km north and south to Hallstatt. In addition the climatic and geographic situation of the region are badly suited for agricultural activities. The oldest salt mining activities are dated to the Middle Bronze Age. Dendrochronology fixes the Bronze Age mining phase to 1458–1245 BC (Grabner et al. 2006).

The actual state of research indicates that three huge shafts systems (depths up to 170 m) operated in parallel (Barth and Neubauer 1991). The enormous amount of archaeological finds and the perfect conditions of preservation in the mines due to the conserving faculties of salt allow for a reconstruction of the work process in the mining galleries (as mentioned in Barth 1993/1994, p. 28). All organic material left in the prehistoric mines has been conserved undamaged due to the preserving faculties of salt (mine timber, wooden tools, strings of grass and bast, hide, fur, textiles, human excrements etc.). This mine waste—also called heathen rock—was left in the mines and has been compressed to solid rock through mountain pressure. The excavated archaeological material from the mines represents almost exclusively tools (e.g. pick handles, collecting tools, carrying buckets) and work assets (e.g. lightning chips, mine timber). Three major areas of Bronze Age mining activity are known, vertical shaft systems can be reconstructed. Salt was mined with bronze picks, producing small pieces of salt, which were then collected with a scraper and trough (see Fig. 7.5) and filled into carrying buckets. These were then carried to the shaft and hoisted to the surface using a wool sack or cloth attached to a linden bast rope. It is assumed that salt was mined on several levels in one mining gallery. The data on mining technology and working processes is dense and of high quality. However, important information is lacking, as no settlement and no cemetery pertaining to the Bronze Age mining phase is known. However, we do have hypotheses based on Bronze Age mining that are derived from anthropological investigations of the Early Iron Age cemetery (9th–4th cent. BC) in the Salzberg Valley. In more detail, the anthropological analysis of the musculoskeletal markers of the excavated skeletons indicates a high workload and specialization on a rather limited range of movements that were iterated over a time span of many years (Pany 2005). The reconstructed movement patterns fit in well with activities related to mining such as breaking salt with a pick and carrying heavy loads. Gender related work division was clearly practiced. Working patterns observed in all studied samples exclude work tasks related to agricultural activities. The anthropological analysis has shown that the age and gender structure of the cemetery correlate with age and gender distributions of a "normal village". Summing up, our inferred hypotheses conclude that (1) Bronze Age miners were working "full-time" in the mine (2) all members of the mining community were involved in the mining process and, in consequence, (3) other groups had to provide them with means of subsistence (food, clothing), (4) the mining community lived in the Salzberg valley.

7.3 Agent-Based Simulation of Mining

7.3.1 Previous Work

The idea of using computer simulations in archaeological research has been around for nearly half a century. The 1970s saw considerable enthusiasm which was then thwarted by the limitations of contemporaneous computer technology and the lack of a sufficiently sophisticated theoretical framework. The developments in computer technology and scientific theory (complex systems theory) in the 1990s have given the application of computer-based modelling to archaeological research a considerable new boost (e.g. Kohler and van der Leeuw 2007, pp. 1–12; Costopoulos and Lake 2010). Especially Agent-based Modelling (ABM) has been popular with the scientific community since the late 1990s. It has been applied to a multitude of research topics, from the development of social complexity, decision-making, culture change, and spatial processes (Doran et al. 1994; Dean et al. 1999; Beekman and Baden 2005; Premo et al. 2005; Clark and Hagemeister 2006) to the exploration of civil violence in the Roman World (Graham 2009) and the work flow analysis in prehistoric mines (Kowarik et al. 2010). What makes ABM especially attractive to archaeology is its potential to model social phenomena on a very advanced level. The bottom-up approach inherent to ABM enables researchers to address individual actions and emergence, thus truly dealing with the complex behaviour of social systems (Premo et al. 2005). The simulation which we are going to present in the forthcoming sections have also previously been reported in Kowarik et al. (2012), in which additional details beyond the scope of this chapter are given. Furthermore, our simulation models are available online http://www.iemar.tuwien. ac.at/processviz/hallstatt as well as upon request to the authors.

7.3.2 Simulating Work Processes

Work processes are given as sequences of actions which are executed repeatedly in order to reach a set goal (in our case: the production of salt). Our agents have no freedom of choice over these strictly defined actions, but execute them as-is. Methodologically, this approach is significantly different from other models that focus on *behaviour*, using a set of rules from which an agent chooses freely. The main difference is that we look at whole work processes as stated by archaeological model building, not emergent behaviour that occurs when agents interact (according to some hypothesis). Arguably, this way of modelling is rather "Tayloristic"; the reason for employing an ABS rather than performing hand-calculation (i.e. "time needed per m^3 of mined material") is that there are dynamic factors which make the result not easily computable lest simulation is used: for example, salt distribution is varying over the simulated area. This can easily be expressed as spatial property 'salt density', which can either be imported or generated, since the typical concentration and form of salt bands are typically known for a specific mountain. Another factor that is dynamic and easily expressed in ABS is the division of work load between different process roles: This may be fixed, or it may vary according to some preset condition (such as a staff schedule, for example).

Our model uses discretized space, in which each cell corresponds to a spot of $1\,m \times 1\,m$ within a mining gallery. By using an additional property, we introduce an additional height of that patch, in multiples of 2 m. Such a *constraint* also applies to the maximum dimension of the mining gallery, which is set to $100\,m \times 40\,m$, 18 m height. Each patch also carries a property that states its salt density, as percentage of salt versus other material present (and which is just 'garbage' in the context of the process). Typically, this density is around 80 %, in a cloud-like shape that can be generated by using a e.g. fractal noise filter peaking at that value. We load this salt distribution as a raster map, i.e. we use a two-dimensional distribution even though our environment is three-dimensional. This approach is nevertheless sound, as salt progresses in vertical bands through the mountain; given this circumstance, we may assume salt distribution to be constant among all layers for the small area in which our model simulates. In other mining simulations which do not satisfy this assumption, the distribution would need to be loaded per layer (i.e. as a set of bitmaps).

As to the simulated personnel, we distinguish between two different agent types (*process roles*): The miner (who breaks the salt) and the transporter (who is responsible for moving the product to a vertical shaft). Current archeological finds suggest a mining process progressing in levels: Each was 2 m high and could be reached from the level below. Furthermore, a spot on the same level could be reached directly, without having to climb up or down. Formalized, this means: a mining spot (see Fig. 7.3) must be at least two cells wide: one for the actual miner agent, one for transporter and reachability. This constraint can be given up if it is assumed that digging and transport take place in sequence.

The maximum number of levels in which the mining is conducted is given as a parameter (the absolute maximum, which comes from the stated maximum dimensions, are nine steps). Initially, we carve out an area of $3\,m \times 3\,m$, which represents the space taken by a vertical access shaft. Furthermore, the first row of patches on each layer are carved out, giving the miners a place to stand on. This initialisation gives a minor deviation from a simulation that would calculate everything from the begin on (i.e. establishing a vertical shaft and the first row

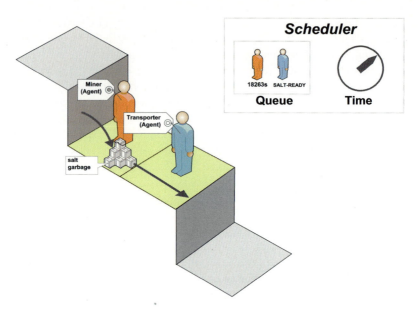

Fig. 7.3 Model of the simulated digging process: miners dig the salt, which is then transported by an own person group to the vertical shaft. Because of the large timespan that has to be simulated (250 years), we have implemented a scheduler that can passivate agents and activate them again when a certain point in time is reached or a specific wake-up signal is broadcast

of patches in each level). The reason for doing so lies in the lack of evidence that would underpin this "mine establishment" part. A detailed discussion of the resulting tradeoff in accuracy is given in the discussion (see Sect. 7.6).

With the levels in place, the agents are now put to work: Miners first query the bottommost layer for all possible mining spots. Mining spots are unoccupied cells which lie beneath a cell on a higher level (i.e. one that has the rock which is to be digged) that contains salt and is also unoccupied. If it has found any such spots, the agent selects the one with the maximum salt concentration, assumes a standing position beneath it and begins digging. If there are no such spots at the current layer, the search advances to the next layer. If all layers are exhausted, the simulation ends.

After having selected a new mining spot, the agents start the actual digging work. Classically, the time base for ABS are ticks of a virtual clock, where each tick stands for equidistant unit of time (e.g. second, hour, year). A unit of work—in our case: the actual digging, takes days (see Sect. 7.5.3). Other processes, such as transport, take only tens of seconds. We have implemented a scheduler *on top of* the ABS, that is made to progress time while the agent is working, which helps us deal with both time resolutions:

- Upon reaching a time-dependent task, the agent is made passive and written into the so-called *future event list* of the scheduler. This list contains all agents that are waiting for a specific instance in time, sorted according to the nearest future activation time.

- Instead of advancing by a second in each turn of the simulation, the scheduler is given control of time. It removes the first agent from its future event list, reads its activation time and advances the clock to that timestamp. Then, it reactivates the agent. It repeats this process while there are agents waiting for the current timestamp.

The scheduler therefore advances time in non-equidistant intervals. For miners, we can specify if the digging should be performed "one whole cell at a time", which means that the scheduler will passivate the agent for the whole amount of time it takes to dig $2\,m^3$ of mountain rock. We can also specify that the time that the miner uses is proportional to the capacity of a carrying bucket, which will be used to transport the salt up to the surface. Regardless of what of the two modes is used, the reactivated agent will mine the amount given by the time passed, taking the salt density (given in percent of salt per cell) into account. It also raises a signal that salt is now available for transport, which will be explained in due course.

Both materials—salt and "garbage" (impure mined rock) increase their volume upon being mined. The increase in volume is fairly significant (in our case $+70\,\%$) for the transport process: There is a special type of agent which is called "transporter", which has the sole duty of collecting the mined salt once the miner agents signal that it is time to do so. For this to be possible, we have extended the scheduler with an additional list, the *future signals list*, that can be used to passivate agents and reactivate them when a certain signal is issued. In more detail:

- Transporters are created, immediately passivated and written to the future signals list in order to wait for *SALT_READY*.
- Miners raise *SALT_READY* after having produced salt. As a matter of fact, the scheduler reactivates all transporters waiting for that signal. The activated agents then go to find salt, one bucket at a time. This means: finding a spot with salt, filling a bucket, transporting this to the vertical shaft. Additional processes, such as the actual transport to the surface and further, are not modelled: We restrict the simulation to the actual process within one mining gallery, in order to make it composable with a simulation of the whole mining facility.

7.4 Implementation

We have implemented the stated model using NetLogo (Wilensky 1999), an open-source simulation that runs on the Java Virtual Machine and is therefore available over a wide range of platforms.

Netlogo itself is a dynamic scripting language, meaning that every command can be issued at run-time (without compilation). Since NetLogo does not support Coroutines (i.e. passivation and reactivation of a piece of code), we had to emulate them using functions: All code that is to be run without interruption is put into a function (or 'Procedure', in NetLogo lingo). At the end of the execution of the function comes the code to passivate the agent. This sets the agent property *active* to false and sets its *resume function*. The agent is then written into either queue (future event or signal list, as described earlier), which is then sorted. The implementation of the future event list uses a priority queue to do that, as the timestamps give the priority.

In every step of the simulation (Fig. 7.4), the scheduler is called and advances its clock to the time of the first item in the future even list. It then takes this agent out of the list, sets its property *active* to true and executes the *resume function*. While there are agents waiting for the same timestamp, this is repeated. The result of this is a set of active agents, which then execute their resume function. Note that, for newly created agents, this is always set to be the first action of the process (e.g. find salt in the case of miners).

Active agents then execute their resume function, which will (1) raise signals such as *SALT_READY*, which are written into a buffer, and (2) also call upon the scheduler to passivate the agent again when a certain time is reached. After that, all agents waiting for a signal are reactivated: For every item in the signal buffer, corresponding agents in the signals list are made active and immediately execute their resume function.

With the help of these constructs, we were able to let an agent simulation compute 200 years of mining in just 5 min (depending on the number of levels). NetLogo's parameter sweeping implementation (*BehaviourSpace*) furthermore allowed us to vary the input parameters and run experiments in unattended mode (e.g. over night).

Fig. 7.4 Actual implementation of the digging process, in NetLogo. The salt distribution is given by the white areas of the world, which the miners follow. At the edges, colour-coded steps are depicted (here for three layers). Both salt production and workforce activities are given as plots. Initial assumptions (constants) are shown on the *right side*

7.5   Experimentation

7.5.1   Parametrization

The simulation was used to obtain the time it would take the agents to fully exploit a mining gallery, given a fixed salt map as basis.

In a pre-step, timing experiments for the described work processes were made inside the mine, using reconstructed bronze-age tools as means. The average of the timings obtained are shown in Fig. 7.5. Furthermore, our simulation used the following parameters as input:

- *Number of layers* (1–9)
- *Number of miners* (25, 50, 100), *number of transporters* (5, 10, 25, 50)
- *Minimal standing space*—number of cells to stand on for each level (1 cell: mining, then transport, 2 cells: mining and transport in parallel)
- *Volume of salt to dig without interruption* (0.02 m^3: one carrying bucket, 2 m^3: one complete cell). This corresponds to the passivation time of the agent, and depends on the salt concentration of the current mining spot.

As maximum simulation time, 250 years were set. This figure is larger than the actual usage time, as dendrochronology gives us 213 years between the youngest and the oldest wood found inside the prehistoric mine. The higher number is considered a safe margin, though.

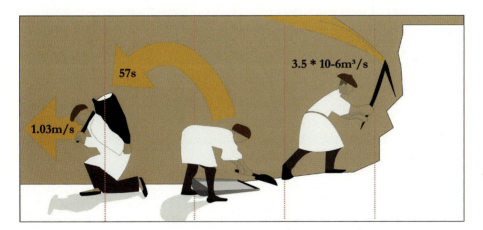

Fig. 7.5 Timings have been derived from experiments which were conducted in the mine, using reconstructed bronze-age tools

7.5.2 Pre-experiments for Narrowing the Parameter Space

The parameter space stated above had to be narrowed down before conducting the parameter sweep experiments:

- Number of layers: Using spreadsheet-calculation, we first obtained the projected total time (number of cells × time per cell), which is dependent on the number of layers that are due to be mined and the minimal standing space. No option could be eliminated, to the contrary: With 25 miners, we calculated 5 years (1 layer) to 39 years (9 layers)—a very small number indeed. It must be noted, however, that these calculations do not take the topological rules stated earlier into account, and are thus the absolute lowest bound.

- Number of miners and transporters: Both of these agents require space. Thus, it is quite possible in initial situations of the simulation that there are too many of them to fit into the mining gallery. The simulation was extended to make non-fitting agents inactive until a space becomes available. Concerning the number of agents, our initial tests confirmed that the time until full exploitation was rather low even when taking topology into account, for example (3 layers, 25 miners digging $2 \, \text{m}^3$ at a time, 5 transporters) 13.88 years of work time. As no constraints concerning the digging workforce were in place, we sticked to the default range (25, 50, 100). It must be noted that a mine that stands still is never 'idle'—even it is fully exploited, since it can still serve as a transit space for horizontal and vertical traffic that has to be maintained through refurbishment of the timber supporting structures.

- The number of transporters was rather insignificant(!) for the total exploitation time. Upon looking closer, the volume of salt that was constantly produced is too small for having a real need to employ distinct transporters. Pictorially speaking, it would have been enough if every miner took the produced salt with him, upon ending his shift. The number of transporters was therefore kept at the minimum (5).

- With the discovery that the transporters were insignificant, it was clear that the minimal standing space could be limited to 1 instead of 2 cells, since parallel transport and mining are rather unrealistic without a surplus of salt at the standing spot of the miner that needs to be brought away.

- The unit of work for the miner did not affect the outcome of the simulation (2.5 % delta for 3 layers, 25 workers, 5 miners). It had, however, a large impact on performance: For $0.02 \, \text{m}^3$ digging at a time, one simulation experiment would compute in the order of hours, whereas the digging of $2 \, \text{m}^3$ would need only minutes (both depending on the number of levels). Therefore, we used $2 \, \text{m}^3$ of digging throughout the simulation. The lack of significance in having a small unit of work is clear when we look at the time it takes to mine a bucket full of salt versus the transport: Assuming 80 % salt density, the miner would interrupt after 1.5 h. The called transporter would need a few minutes to scrape the salt together, load it into a bucket and bring it to the shaft—before becoming idle again. On the contrary, having $2 \, \text{m}^3$ of salt mined in one piece requires 1.7 h of work for five transporters (assuming a transport takes 3 min in average), which is still insignificant. As a note of caution, we once again state that all of these considerations of transport do not take the vertical lifting from the shaft to the surface into account, which is currently being researched using physical simulation.

7.5.3 Experiments

In the actual experiments we simulated the reduced parameter space that we got as result of the pre-experiments. We used 10 repetitions per for layers 1–7, and one run per parameter variation of layers 8 and 9. As justification, it is noted that there was no variation in the produced values of layers 1–7 (<0.01 years), and the required time for a run in levels 8 and 9 was very large (>1 h).

7.5.3.1 Quantitative Results

We could show that the number of people actively working in the mining gallery might have been smaller than initially assumed, given the necessary time for totally exhausting a mining gallery. Depending on the number of levels (see Fig. 7.6), 25 workers need between 4.74 and 41.39 years. Adding more workers does not decrease the work time in a directly proportional manner: For example, doubling the 25 workers found does not decrease the needed time to 50 %, but only to 54 %.

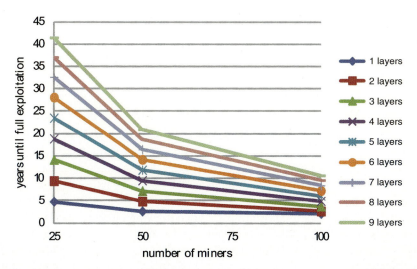

Fig. 7.6 Results of our mining model. Time until full exploitation (y axis) vs. number of workers (x axis) vs. number of mining galleries ("layers", see graph lines)

Much of this effect is caused by the limited space in the initial stages of the simulation, where not every agent can be fitted inside the given space and must therefore stay idle (waiting in the vertical access shaft), along with the obvious different mining pattern occurring if more agents are at work.

7.5.3.2 Qualitative Results

The shape of the mining galleries produced can be seen as *emergent outcome*: In contrast to the initial idea that mining would produce large mining galleries with carved-out levels along the boundaries, we observed a multitude of connected halls with smaller connecting hallways. The agents tend to dig on one level in a sweep-like motion, before breaking creating a hallway and passing through into the next mining gallery. We interpret this to be the caused by the shape of the salt concentration map, in which the salt distribution is "cloud-like" (as in reality) and the behavioral rule to focus on the piece of rock with the highest amount of salt. Recently another possibility has entered the archaeological discussion: It is now assumed that the galleries were mined according to a preconceived plan: Based on different findings from the excavations in the salt mine, it seems to emerge that the prehistoric miners did not simply start mining wherever they found the highest content in rock salt. It rather seems that already before the mining started a "construction plan" for the mine existed or to put it differently the shape of the mining halls was fixed even before they came into existence.

As a further qualitative outcome, we found that the idea of having an extra walkway of at least one patch in each level makes sense only if the mining work must proceed highly parallel between miners and transporters, which is doubtful at least when looking at the production rate for salt. As transporters are rather insignificant given our result, this constraint does really not seem to apply, if not other considerations than parallelism come into play.

7.6   Discussion

Our simulation computes pure working times, in years. We have no social model or other forms of time constraints that govern our miners behind our model. Even if we did try to apply such a mechanism, e.g. 8 h for work, 8 h free time and 8 h sleeping, the time until full exploitation would be far smaller than the actual period of use for the mine (213 years). Some questions are therefore: Was the mining hall exhausted and then only used for accessing deeper-lying mining galleries? Or, was salt production on such a small scale that it would fill the whole time-span?

When considering our simulation results in further consequence, we also find that they sometimes contradict longstanding archaeological assumptions. For example, bronze picks, salt buckets and scraping tools were thought to be designed to optimize the working process, as they both consist of standardized parts and were intended for performing the same work steps over and over in rapid succession. However, the simulation suggests that such a kind of efficiency was not needed, due to the slow rate of salt production.

What have we left out? A specific trail that we did not follow is the change in the spatial environment, which is assumed to be static in our model. Leftover debris and burnt-down torches effectively alter the shape of a mining hall to such an extent that movable (wooden) staircases had to be put on the rubble to establish walkways. Clearly, simulating and verifying the distribution of the rubble nowadays found in the prehistoric mine would be an interesting work for the future. It does, however, not change the overall exploitation time, which we have simulated here.

Another shortcoming, which we have accepted because of the lack of data and consequently also of experimentation results, is the way in which the vertical shafts were built. It is clear from the excavated parts of the mine that a certain amount of timber constructions were installed in the vertical shaft. However, a large part of reconstructing these shafts needs to be based on the lifting mechanisms, which we are still trying to understand. Clearly, linden bast ropes were used to lift weights; the actual lifting construction has been not excavated to its full extent, and therefore, the surrounding staircases and/or ladders need to be clarified. Some preliminary results point at the weight of the lift itself, which might have been very large considering the frictional forces and the weight of the rope itself. Several alternatives—a closed-loop rope versus an open one—are still being researched, which is why we do not seek to jump to conclusions at this early stage. The vertical circulation represents one of the most challenging issues that we are faced with, which is going to occupy us for the next few years further on the research trail. Current efforts in that context are: The physical simulation of the lifting mechanism, the staircase construction (based on the wood distribution coming from the actual, collapsed staircase) and the research into the collapse itself.

7.7 Extensions: Towards Prehistoric Economy

As augmentation of the basic mining model, we have developed a mixed system dynamics/ABS model for simulating demographic development in the Hallstatt region (see Fig. 7.7). The basic question was: If the results from the presented mining simulation cannot be used to narrow down on the number of people working simultaneously, demographic development can perhaps give a measure on how many people might have been available.

System Dynamics (SD) is a method for simulating complex systems with nonlinear behaviour, roughly consisting of two conceptual parts:

- *Stocks* holding a numeric quantity (of people, in our case)
- *Flows* connecting to and leading away from stocks, used for controlling how much of the quantity enters and leaves a stock in each time unit (which is continuous, in contrast to ABS).

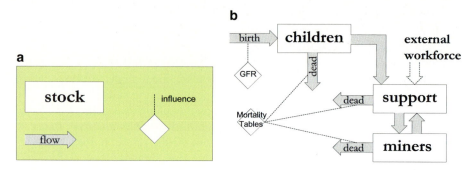

Fig. 7.7 (**a**) System Dynamics concepts: stocks containing quantities, flows modeling the change in quantities, variables giving influences on flows and stocks. (**b**) SD model used for the demographic simulation

The model is entered as graph: Stocks are being shown as nodes and flows as edges (see Fig. 7.7a). In our SD extension, the stocks stand for different stages population, i.e. non-working children, support workers, miners, and time of the simulation naturally corresponds to growing up. Since SD has no notion of identity for the contents of a stock, and thereby no way of telling "how many persons of what age" are contained, we have used a coupled ABS to actually represent the individuals of a population. This way, we can apply age-based mortality rates according excavation data of the Early Bronze Age cemetery of Franzhausen (Berner 1992). For births, we used a fairly high General Fertility Rate (GFR, defined by WHO as "number of live births during a year per 1000 female population aged 15–49 years (reproductive age group)") starting with 200 as base number. Such a high number is found nowadays in developing countries (see e.g. United Nations 2009), which is an analogy we wish to draw.

In each step of the simulation, the SD computes population continuously in terms of *changerate* * dt, where each unit of time stands for a year and the changerate corresponds to the time-dependent flow. In contrast to that, the coupled ABS does only know of integer quantities. Therefore, changes have to be rounded down (e.g. 20.2 births yield 20 agents in the ABS, the remaining 0.2 are accounted for in the next cycle of the SD).

The actual population simulation is summarized as follows (refer to Fig. 7.7b):

- We start by adding births to the stock of children (flow birth – children), where birth depends on the GFR. Simultaneously, agents of age 0 are created in the linked ABS.
- The progression from the children stock to the stock containing people doing support work (i.e. juveniles, old people, people for whom there is currently no employment in the mines) is controlled by age (flow children – support). Furthermore, mortality is given by the flow children – dead is calculated, as given by mortality for every age class.

- If needed, people in the support stock will be assigned to the miners stock, or released from the latter into support (flow support – miners). As always, mortality is also modeled by the flow support – dead, miners – dead.

We aimed to answer the question of how high the General Fertility Rate needed to be in order to sustain a population fit to provide a fixed set of miners (e.g. 5, 10, 15, 20) over a longer time period (here: 300 years). The first simulation runs confronted us with the problem that we had to choose a rather big initial population and considerable fertility rates to sustain a stable population over 300 years. Therefore we focus here on the necessary parameters for sustaining a stable population.

Simulation runs were conducted for a GFR range from 200 to 425, 50 experiments per parameter variation. We worked with an initial population of 300 people. Beginning at a GFR of 375 (every woman between 15 and 49 needs to have a live birth every 2.5 years), stable conditions over a time span of 300 years are obtained (see also Fig. 7.8). The simulation was then expanded to integrate migration to Hallstatt.

Migration is characterized by the addition of people into the support class (age between 10 and 16). These people have already passed the initial hurdles (most noteworthy: 58 % mortality in between 0 and 4 years) and are on the edge of their reproductive age. Figure 7.9 shows the simulation results of such a population dynamics (GFR 225 is assumed):

- An estimated migration of 4.75 persons per year sustains the initial population.
- However, even a lower migration of 1 person per year establishes a stable population of 75 persons, which would probably be in the range that allows to run a mining facility.

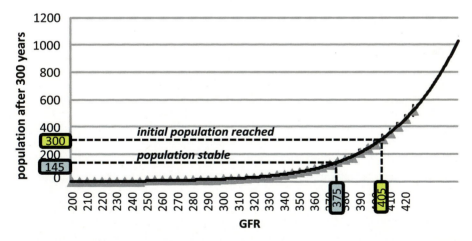

Fig. 7.8 Sustainability of the population in Hallstatt when only GFR is taken into account. A GFR of 375 would lead to the stabilization of the population on a very limited level, a further increase would produce the initial population of 300

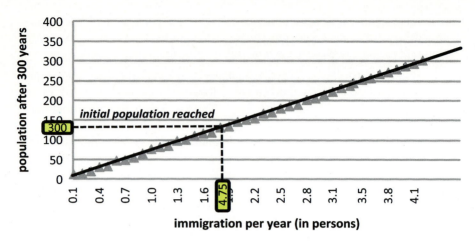

Fig. 7.9 After introducing immigration and leaving the GFR at 225, the initial population is reached quite easily. This is due to the fact that immigrants have already lived through their childhood, which has a high associated mortality

Several aspects need to be taken into account before discussing this output any further:

- Mortality rates were taken from an early Bronze Age cemetery in the lowlands of Eastern Austria (Franzhausen/Lower Austria, according to Berner 1992).
- The demographic model is based on a rather simple structure not taking into account illness, warfare, natural catastrophes. The model needs to be reevaluated as it represents a first trial version.

Keeping these points in mind two possible scenarios emerge under the given model constraints (also see Sect. 7.2):

- a local population without migration, but, in our view, high General Fertility Rates
- a local population with a certain amount of migration to Hallstatt and somewhat lower fertility rates

Looking at the archaeological and anthropological record arguments can be made for both scenarios.

Concluding, the demographic simulation provides important "food for thought" introducing demography as a possible important limiting factor in the operation of the salt mines.

A further perspective would then be to model the whole support community versus the actual workforce occupied in digging. As in modern-day business engineering, we have used a business process simulation (see Fig. 7.10) to model multiple levels of mining activities, the tools needed and salt produced. The model is still in its early stages, solely concentrating on the mine as a system which integrates the presented simulations, in order to answer on demand and supply levels. Still, it is

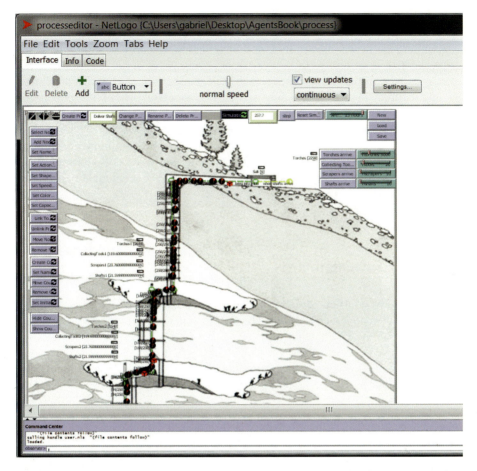

Fig. 7.10 A business process simulation can simulate the prehistoric mine from a systems view, i.e. amount of materials needed for producing the needed quantities of salt, just like in a current enterprise that has to optimize its daily work procedures. It is arguable, though, if the prehistoric mine was set up in exactly this spirit

to early to give decisive results on this simulation, whose role is to visualize the different parts coming together, and ask about probable exploitation strategies (one mining gallery at a time, or multiple mining galleries in parallel).

7.8 Conclusions and Future Work

We have presented a mining model that acts on the basis of the reconstructed work process for digging salt in the bronze-age mines of Hallstatt/Upper Austria. Our results obtained so far seem promising: We could show that (1) the work force needed for exploiting such a mine would have been smaller than previously assumed and (2) that it might have been a severely limiting factor in the mining system.

As future work, we need to look at the vertical transportation of the salt to the surface, which is currently being investigated using physical models of the lifting process and the vertical shaft construction itself. Furthermore, we have a huge need for investigation of the surrounding settlement: Sustainability of the population (based on landscape properties), also including migration, trade of goods, social models of a mining community and the like will keep us occupied for the coming years, to say the least. All in all, the contribution of simulation in this respect lie not solely in the actual results: The formalization of verbal models alone augment the daily archeological practice in ways that were previously unthought of, and will continue to inform other scientific disciplines also involved in the research in ways that are yet neither imaginable nor foreseeable.

Authorship Information

All authors have contributed equally to this chapter. As head of the excavations in the prehistoric mines of Hallstatt, Hans Reschreiter has provided the necessary background that the whole team then formalized into a simulation model. Kerstin Kowarik has played a key role in this formalization, by filling in the gaps during the constant back and forth between current state of research and the possibilities of the simulation, which Gabriel Wurzer implemented in Netlogo.

Acknowledgements The excavations in Hallstatt have reached a scale at which research cannot be made in isolation; we are grateful for a large body of institutions and people supporting the work that is being conducted, among which we want to explicitly name the Natural History Museum Vienna (especially the Prehistoric Department), the Salinen Austria AG, the Vienna University of Technology (Department of Architectural Sciences, Department of Analysis and Scientific Computing), the University of Natural Resources and Life Sciences Vienna (Department of Material Sciences and Process Engineering) and the Vienna University (Institute of Prehistoric and Historical Archaeology). From the on-site staff, we would especially like to thank Ralf Totschnig for conducting the experiments and Andreas W. Rausch for his documentation. Furthermore, we are grateful for the earlier work conducted by Fritz-Eckart Barth, which has paved the way for all of the current research activities that we undertake.

References

Barth FE (1993/1994) Ein Füllort des 12. Jahrhunderts v. chr. im Hallstätter Salzberg. Mitteilungen der Anthropologischen Gesellschaft Wien 123/124:27–38
Barth FE, Neubauer W (1991) Salzbergwerk Hallstatt. Appoldwerk, Grabung 1879/80. Salinen Austria, Verlag des Musealvereines Hallstatt, Hallstatt
Beekman CS, Baden WW (eds) (2005) Nonlinear Models for Archaeology and Anthropology - Continuing the Revolution. Ashgate, Hampshire-Burlington

Berner M (1992) Das frühzeitliche Gräberfeld von Franzhausen I, Niederösterreich - metrische und demographische Analyse. PhD thesis, University of Vienna

Clark J, Hagemeister EM (2006) Digital Discovery: Exploring New Frontiers in Human Heritage. In: Clark D, Clark J, Hagemeister EM (eds) Proceedings of the 34th Conference on Computer Applications and Quantitative Methods in Archaeology. Archaeolingua, Fargo, pp 16–59

Costopoulos A, Lake M (2010) Introduction. In: Costopoulos A, Lake M (eds) Simulating Change: Archaeology into the Twenty-First Century. University of Utah Press, Salt Lake City

Dean JS, Gumerman GJ, Epstein JM, Axtell RL, Swedlund AC, Parker MT, McCarroll S (1999) Understanding Anasazi Culture Change Through Agent-Based Modeling. In: Kohler T, Gummerman GJ (eds) Dynamics in Human and Primate Societies. Agent-Based Modeling of Social and Spatial Processes. Oxford University Press, New York, pp 179–207

Doran J, Palmer M, Gilbert N, Mellars P (1994) The EOS Project: Modeling Upper Paleolithic Social Change. In: Doran J, Gilbert N (eds) Artificial Societies. UCL Press, London, pp 195–221

Grabner M, Reschreiter H, Barth FE, Klein A, Geihofer D, Wimmer R (2006) Die Dendrochronologie in Hallstatt. Archäologie Österreichs 17(1):49–58

Graham S (2009) Behaviour Space: Simulating Roman Social Life and Civil Violence. Digital Studies/Le champ numérique 1(2). http://www.digitalstudies.org/ojs/index.php/digital_studies/article/view/172/214

Kohler TA, van der Leeuw SE (2007) Introduction: Historical Socionatural Systems and Models. In: Kohler TA, van der Leeuw SE (eds) The Model-Based Archaeology of Socionatural Systems. School for Advanced Research Press, Santa Fe, pp 1–12

Kowarik K, Reschreiter H, Wurzer G (2010) Modeling A Mine: Agent Based Modeling, System Dynamics and Experimental Archaeology Applied to the Bronze Age Saltmines of Hallstatt. In: Proceedings of the 1st Mining in European History-Conference of the SFB HiMAT

Kowarik K, Reschreiter H, Wurzer G (2012) Modelling prehistoric mining. Math Model 7(1): 17–29. doi:10.3182/20120215-3-AT-3016.00005. iFAC-PapersOnLine

Pany D (2005) Working in a saltmine... Erste Ergebnisse der anthropologischen Auswertung von Muskelmarken an den menschlichen Skeletten aus dem Gräberfeld Hallstatt. In: Karl R, Leskovar J (eds) Interpretierte Eisenzeiten. Fallstudien, Methoden, Theorie. Tagungsbeiträge der 1. Linzer Gespräche zur interpretativen Eisenzeitarchäologie. Studien zur Kulturgeschichte von Oberösterreich, Landesmuseum Linz, vol 18

Premo LS, Murphy JT, Scholnick JB, Gabler BM, Beaver JE (2005) Making a case for agent-based modeling. Soc Archaeol Sci Bull Fall 2005 28(3):11–12

United Nations (2009) World Fertility Report 2009. Department of Economic and Social Affairs, Population Division

Wilensky U (1999) NetLogo. Center for Connected Learning and Computer-Based Modeling, Northwestern University, Evanston, IL. http://ccl.northwestern.edu/netlogo/

Chapter 8
Modelling Settlement Rank-Size Fluctuations

Enrico R. Crema

8.1 Introduction

This chapter explores the underlying causes of changes in settlement rank-size distribution by modelling the dynamics of group fission and fusion and their responses to different disturbance regimes. The theoretical framework underpinning this exercise is based on the following assumptions:

- The amount of resources at a given location can influence the size of a group located there;
- The relationship between group size and per-capita fitness is expected to increase with increasing group size. Once a critical threshold is exceeded, this relationship is reversed;
- Individuals are expected to improve their condition by means of spatial repositioning, though this will be constrained by limits in knowledge and energy.

An agent-based simulation has been developed in order to establish how variations in the details of these assumptions can induce divergence in the system equilibria, and then to explore how different forms of perturbations (mimicking various forms of endogenous and exogenous environmental deterioration) can alter these.

The chapter will be structured as follows: Sect. 8.2 will provide the background discussion, including an overview on some of the theories underpinning the proposed model; Sect. 8.3 will discuss the details of the agent-based model and how the three assumptions listed above have been formalised. It will also introduce the four different models of disturbance processes examined here; Sect. 8.4 will present the experiment design and the results of the simulation exercise; Finally, Sect. 8.5

E.R. Crema (✉)
University College London, London, UK
e-mail: e.crema@ucl.ac.uk

© Springer International Publishing Switzerland 2015   161
G. Wurzer et al. (eds.), *Agent-based Modeling and Simulation in Archaeology*,
Advances in Geographic Information Science, DOI 10.1007/978-3-319-00008-4_8

will discuss the wider implications of the model and the main conclusions of the
chapter.

8.2 Background

Fission and fusion of human groups can be inferred for a wide variety of temporal
scales. Intra-annual events are ethnographically known for many hunter-gatherer
groups, who often aggregate temporarily into large groups, only to disperse soon
after. For example, the Nootka Indians of the Pacific Northwest coast aggregated
into large confederacy sites during the summer while they fissioned into smaller villages during the winter (Drucker 1951, cited in Watanabe 1986). Other ethnographic
evidence shows how these fission-fusion cycles can occur with much less regularity
and lower temporal frequencies. Historical census data from the Hokkaido Ainu
hunter-gatherers provides a good example in this regard. During an interval of 14
years, several sedentary households of the Mitsuishi district fissioned from larger
groups or formed new settlements in an irregular fashion (Endo 1995). At a further
larger temporal scale, the alternation between dispersion and nucleation of farming
communities (Roberts 1996; Jones 2010) have been detected from both historical
and archaeological evidence.

Variations in the settlement size distribution are ultimately the result of two
processes: the *movements of individuals* and inter-group *differences in the intrinsic
growth rate*. The two are related to each other, and in most cases available
archaeological evidence is not sufficient to distinguish the outcome of one from
the other. However, we can acknowledge their existence if we identify variations
in the residential density of a region (a cumulative effect of changes in the overall
growth rate) or if we detect the presence of newly formed settlements in a given
time window (a direct consequence of fission events).

Despite the difficulty in obtaining direct and reliable proxies of settlement sizes,
archaeologists have been long interested in measuring the temporal variation of
settlement hierarchy, one of the most tangible consequences of these processes.
However, the skewed and long-tailed shape of most settlement size distributions
makes the adoption of common statistical measures impractical. Hence, settlement
systems are often described using the relationship between rank and size formalised
in the following equation (Zipf 1949):

$$S_r = S_1 \cdot r^{-q} \tag{8.1}$$

where S_r is the size of the r ranked settlement, and q is a constant. Equation (8.1)
establishes a power-law relationship between size and rank, where the slope is
defined by q. When this constant is equal to 1, we obtain the so-called Zipfian
distribution, originally considered as equilibrium between "forces of unification"
and "forces of diversification" (Zipf 1949).

Fitting equation (8.1) to archaeological data and obtaining empirical estimates of q is a straightforward exercise, and allows us to quantitatively classify different settlement systems. Thus, we can refer to *primate systems* when $q > 1$, that is when we have few large and many smaller settlements. Conversely, when $q < 1$, the system can be classified as *convex*, with the size distribution being more uniform than the Zipfian expectation. However, Drennan and Peterson (2004) noticed that most archaeological data do not appear to conform to such a log-linear relationship between rank and size, and thus devised a more flexible measure explicitly based on the amount of deviation from the theoretical Zipfian distribution ($q = 1$). Their A-coefficient analysis is computed in two steps. First the observed rank-size plot is rescaled, so that the area defined by the end-points of the theoretical Zipf-law distribution is equal to 2. Then the area between the observed and theoretical rank-size distributions is computed, with the area of sections beneath the Zipf's law pattern multiplied by -1. This ensures that the resulting number (the A coefficient) is positive (up to 1) for convex, negative for primate, and close to zero for Zipfian systems. The application of the A-coefficient analysis has increased the number of archaeological cases where the empirical evidence suggests the existence of long-term fluctuations between primate and convex systems (e.g. Drennan and Peterson 2004; Kohler and Varien 2010; Crema 2013a). Several authors have proposed models of generative processes behind these empirically observed rank-size distributions. Hodder (1979) compared the goodness of fit of different stochastic growth models to archaeologically detected rank-size distributions, while more recently Griffin (2011) developed an agent-based model where cycles of consolidation and collapse of complex polities is the primary driver of changes in settlement hierarchy. Others have suggested theoretical linkage between known settlement models and expected deviations from the Zipfian distribution. Thus central place theory, territorial isolation, and low system integration have been linked to convex settlement patterns, while the spatial concentration of resources to the emergence of primate systems (Johnson 1980; Savage 1997).

The two fundamental processes mentioned above (*difference in growth rate* and *movement of individuals*) are still central in these models. Difference in growth rate can be a consequence of variation in resource availability; isolation, low-level integration, and territoriality can be effectively conceived as constraints in the movement of individuals. Here, I consider two sets of theories proposed in behavioural ecology that provide a robust and flexible framework for modelling these two processes.

The first set looks at the attractive and repulsive effect of the external environment, primarily expressed in terms of resource availability. This induced form of spatial dependency (Fortin and Dale 2005), is the central concept of the Ideal Free Distribution (IFD) models (Fretwell and Lucas 1970; Tregenza 1995). The basic prediction in this case is that, given an omniscient population with a complete lack of constraints in movement, the expected population density on a patch will be proportional to the local resource density. This idea, often referred to as "habitat matching rule" (Fagen 1987), is a consequence of an assumption formally described by the following equation:

$$\phi_i = K_j/n_j \qquad (8.2)$$

where the fitness or gain (ϕ) of an individual i at patch j is the ratio between the amount of resource (K) and the number of individuals (n) located there. Thus Eq. (8.2) will be maximised with the lowest population density, and any increase of n will determine a decline in fitness. With other things being equal, individuals will avoid choosing a patch with high resource input if the local population becomes high, and might opt for a patch with lower K as long as n is significantly lower there. This assumption has been further extended, to include the possibility of interference in foraging activities (Sutherland 1983) and the potential to exercise constraints in the movement of other individuals (i.e. ideal despotic distribution; Fretwell and Lucas 1970). Some of these models have also been applied to predict colonisation sequence and settlement history (Kennett et al. 2006; Winterhalder et al. 2010).

One of the key implications of IFD is that aggregation is an indirect consequence of resource distributions. Individuals are "pushed" together, attracted by the presence of richer habitats. Thus a convex settlement pattern should be expected for a landscape with a homogenous distribution of resources while a more primate distribution should result from a heterogeneous setting.

The "push" argument underpinning IFD becomes problematic if one considers the benefits that can potentially derive from aggregation alone. This, in fact, opens to the possibility that individuals might be also "pulled" by the presence of other individuals. Examples of such a positive frequency dependence arising from group formation have been exhaustively discussed in anthropology and ecology, ranging from the benefit of mutual protection (Gould and Yellen 1987) to the possibility of cooperation and more complex organisation of tasks (Hawkes 1992). The presence of these positive frequency dependencies at small population density coupled with negative frequency dependencies at larger sizes is often referred to as the Allee effect in ecology (Allee 1951). The implications of such a non-linear relationship are crucial, and can often lead to unexpected macro-level dynamics. For example, Greene and Stamps (2001) showed how the integration of Allee effect to standard IFD models can lead to the emergence of population clusters that cannot be explained by properties of the resource distribution. Although not explicitly referring to the Allee effect, several authors (Sibly 1983; Clark and Mangel 1986; Giraldeau and Caraco 2000) have also explored the consequences of this non-linear relationship, suggesting, for example, how the expected group size is not necessarily the one in which fitness is maximised (the "optimal group size"), but the one in which this becomes equivalent to the fitness expected by the smallest possible group (the "equilibrium group size").

There are several further assumptions that we need to incorporate in to our model. The foremost is the role of time and, consequently, aspects pertaining inheritance and path dependence in the system of interest (Premo 2010). The Allee effect implies that the attractiveness of a group will dynamically change depending on the decision of other individuals. Small differences emerging from stochastic components in the system could induce migration flows towards a given group, increasing the fitness of its members, and hence provoking a positive feedback loop. In the long term, however, this process is expected to promote the opposite behaviour, as once optimal group size is reached, fitness will start to decline and individuals will do better leaving the group. As a corollary to this, we also need to consider that fitness will directly affect the long-term behaviour of the system in terms of variation in the intrinsic growth rate.

Similarly, we need to take into account that Eq. (8.2) considers K as parameter constant, and hence invariable over time and by the activities of the local population. The standard IFD model assumes that resources are instantaneously regenerating and hence the abandonment of a patch (and the consequent decline in n) will lead immediately to an increase in the fitness of the individuals who remain there. Externally induced changes in the resource input could tilt the equilibrium of a system, and similarly a reciprocal feedback process between resources and individuals (e.g. K varying over time as a function of n in the past) can lead the system to different equilibria.

Lastly, the level of integration between sub-components of the system (individuals, groups, etc.) should be considered. The non-linear relationship between group size and fitness has been primarily explored without considering the implications of multiple groups co-existing in the landscape. Once we add this to the model (e.g. Greene and Stamps 2001), the dynamics will be partly affected by the level of integration between communities, measurable in terms of physical constraints in the movement (i.e. the cost associated in moving from one place to another, frequency of movement, etc.) and knowledge (i.e. where to go).

8.3 Model Design

We can formalise and extend the three assumptions listed in Sect. 8.1 by generating an agent-based model that embraces the theoretical framework discussed so far.

8.3.1 Basic Model

Consider a population of n agents dispersed in a toroidal landscape composed by P patches. We define a group as a subset of the population of agents located in the same patch. The maximum number of groups will thus be P, and each group j will have a size g_j, defined as the number of agents located in the same patch j. The simulation will proceed through a sequence of discrete time steps $t = 1, 2, 3, \ldots, T$ where the distribution $G_t = g_{j=1}, g_2, g_3, \ldots, g_P$ will be updated by two key processes: *intrinsic population growth/decline* of each group (i.e. reproduction and death), and the *movement of the agents*. Notice that G_t will be essentially equivalent to the settlement-size distribution at given moment in time t, and hence can be quantified in terms of rank-size. Here, we chose to use the A-coefficient (Drennan and Peterson 2004) described earlier for its flexibility in describing a wider range of patterns. Thus, for each run of the simulation we generate a time series A_t describing the rank-size dynamics of the system.

The core component of the model, which affects both key processes, is the computation of the agent's fitness ϕ. This will be executed in two steps. First the "demand" ξ_i of each agent i will be computed as a random draw from a normal distribution with mean $\mu + (g-1)^b$ and standard deviation ε, where μ is the basic fitness (i.e. the expected yield without cooperation), b is the benefit derived from cooperation, g is the local population density (i.e. the group size), and ε is the stochastic effect of foraging tasks. With other things being equal, ξ_i will increase linearly with increasing group size. The *Allee* effect will be introduced in the second step of the fitness evaluation with the following equation:

$$\phi = \begin{cases} \frac{\sum_i^g \xi_i}{g} & \text{if } \sum_i^g \xi_i < K \\ \frac{K}{g} & \text{if } \sum_i^g \xi_i \geq K \end{cases} \tag{8.3}$$

where K is the amount of resource available at the local patch.

The relationship between individual fitness and group size could be potentially modelled in several ways (see Clark and Mangel 1986 for other plausible models), but Eq. (8.3) encapsulates some of the core assumptions regarding human aggregations:

- Grouping provides benefits in the per-capita fitness;
- Some of these benefits will decline in their effect with increasing group size;
- With a further increase in group size, negative and detrimental forces will become predominant, with a resulting decline in the per-capita fitness.

These three points characterise the *Allee* effect described in Sect. 8.2. Here, increasing b will determine a higher average mean per-capita fitness, as long as the total "demand" (sum of all ξ) of the group does not exceed the available amount of resources K. In such a case, the positive effect of cooperation will no longer be sufficient, ultimately leading to a decline in fitness (ϕ).

This non-linear relationship becomes a key element once we explore the two sets of processes that modify group sizes: variation in the intrinsic growth rate and movement of the agents. For the former case, we can translate fitness into a net growth rate, defined as the difference between the probability of reproduction (r) and death (d). We can formalise this as follows:

$$r = \rho \frac{\phi}{\mu} \tag{8.4}$$

$$d = (1 + e^{\phi \omega_1 - \omega_2})^{-1} \tag{8.5}$$

Table 8.1 Fission-fusion conditions and agents' decision-making

Condition 1	Condition 2	Decision
$g_i > 1$ AND $g_w > 1$	$\phi_i \leq \mu - c$ AND $[\phi_w \leq \mu - c$ OR $\phi_i \geq \phi_w]$	Fission
	$\phi_w > \mu - c$ AND $[\phi_i \leq \phi_w - c$ OR $\phi_i \leq \mu - c]$	Migration
$g_i > 1$ AND $g_w = 1$	$\phi_i < \mu - c$ OR $\phi_i < \phi_w - c$	Fission
$g_i = 1$ AND $g_w > 1$	$\phi_i \leq \phi_w - c$	Migration
$g_i = 1$ AND $g_w = 1$	$\phi_i < \mu > \phi_w$	Fission
$g_i > 1$ AND $g_w =$ NULL	$\phi_i \leq \mu - c$	Fission

For all other conditions, the agent stays in the patch where it is currently located; Fission = the agent leaves the group and form a new group with size 1, as long as an empty patch is available within distance h; Migration = the agent joins the group of the model agent w; Fusion = the focal and model agent form a group of size 2

Equation (8.4) establishes a linear increase in the reproductive rate of the agents (controlled by ρ), while Eq. (8.5) has a sigmoidal shape with a small probability of death at high values of ϕ, and an exponential increase of mortality at lower values (cf. Pelletier et al. 1993).

The movement of each agent is assumed to be driven by a mixture of "meliorising" and "satisficing" principles (Mithen 1990), where the key element for evaluation is the "perceived" difference in the observed fitness. The model will produce fission and fusion dynamics based on the following algorithm, triggered with frequency z:

1. A focal agent i defines a pool of observed agents S as a random sample of proportion k of agents located within distance h from i.
2. The agent with the highest fitness among the pool S will be defined as the model agent w. If S is an empty set, there will be no model agent.
3. The focal agent i will compare its own group size (g_i) and fitness (ϕ_i) with: the model agent group size (g_w), the model agent fitness (ϕ_w), and the basic fitness (μ). The comparison will be calibrated by a threshold of evidence c (Henrich 2001), representing the propensity of the agent to be conservative (high c) or not (low c).
4. As a result of this comparison the agent will decide to stay in the current group, join another group, or form a new group on its own (see Table 8.1).

8.3.2 Integrating Disturbance

The model presented so far is primarily defined by parameters that describe the behaviour of the agents. The only exception is the resource input size K, a state variable of the patches where the groups are located, and hence independent to the agents. Thus if we want to explore the intrinsic properties of the system we can assume K as a constant and invariable parameter. This can provide a benchmark model (*scenario 0*), where we can identify the key properties of the system in a controlled condition where the dynamics are exclusively the consequence of the agents' behaviour. Subsequently, we can relax this assumption, and explore the effects of disturbance, i.e. variation of K, under the following four scenarios.

The first (*scenario 1*) explores the effect of spatial heterogeneity by adding to the initial homogenous distribution of K a random integer with mean 0 and variance v. Increasing values of v will increase the heterogeneity of the resource distribution, maintaining, on average, the total productivity (the sum of all K of all cells) of the system constant. The benchmark model (*scenario 0*) can be regarded as a special case of this scenario where v is equal to 0.

The second scenario (*scenario 2*) will relax the assumption of the temporal homogeneity, allowing K to be time-variant. This will be modelled as a bounded random walk, iterating the same algorithm used for scenario 1 for each time-step in the simulation, again parameterised by v. To avoid excessively high or low values of K, the process will be "bounded" between K_{lo} and K_{hi}. High values of v will generate abrupt shifts, while lower values will lead to gradual changes in the resource availability. The third scenario (*scenario 3*) will combine the assumptions of scenarios 1 and 2, allowing the resource input of each patch to have an independent bounded time-series of K.

In contrast to the models of disturbance proposed so far, the last scenario (*scenario 4*) shapes the spatio-temporal variation of K as a result of a predator-prey relationship with the agents. The assumption in this case is that high local population density should, in the long term, determine a degradation of the local environment and a decline in resource productivity. This differs somewhat from the detrimental role of overexploitation portrayed in Eq. (8.1), as the effect will be also time-dependent (i.e. a group might experience a long-term decline in fitness even if g is hold constant). The predator-prey relationship can be formalised with the following pair of equations:

$$\varXi = \begin{cases} \sum_i^g \xi_i & \text{if } \sum_i^g \xi_i \leq K_{t-1}(1-\beta) \\ K_{t-1}(1-\beta) & \text{if } \sum_i^g \xi_i > K_{t-1}(1-\beta) \end{cases} \tag{8.6}$$

$$K_t = (K_{t-1} - \varXi) + \zeta(K_{t-1} - \varXi)\left(1 - \frac{K_{t-1} - \varXi}{\kappa}\right) \tag{8.7}$$

Equation (8.6) defines the cumulative gain $\varXi$ of the agents—which becomes the individual fitness once its divided by the group size—and is subtracted from the resource input in Eq. (8.7), a variant of the Verhulst equation (Verhulst 1838), defined by an intrinsic growth rate ζ and a carrying capacity κ. Equation (8.7) thus ensures that K is modelled as a population affected by the consumption rate $\varXi$ of the agents. The parameter β in Eq. (8.6) models the resilience of the resource pool: high values will determine an under-consumption of the agents (i.e. the agent will not be able to identify and consume all resources located on a given patch), while low values will increase the likelihood of complete resource depletion.

8.4 Results

The simulation code was written in R statistical computing language (R Core Team 2013) and is available under request. All experiments have been conducted using UCL Legion High Performance Cluster. A wider exploration of the parameter space for the benchmark model (*scenario 0*) is extensively discussed elsewhere (Crema 2014). Here we purposely sweep only the key parameters that have been previously identified as those[1] determining the largest variation in the system behaviour: the spatial range of interaction h; the frequency of decision-making z; and the sample proportion of the observed agents k. We additionally sweep three values for relevant parameters describing different disturbance processes (v for scenarios 1, 2 and 3, and β for scenario 4). In this case, the choice of parameter values has been dictated by preliminary explorations of the model in a simplified environment with a single group (P = 1), where the effect of movement has been excluded. This exercise allowed the detection of a key range of values covering the widest spectrum of behaviours in the simplified model. For example, a small variation of β from 0.3 to 0.4 was sufficient to cover the phase transition between three equilibria in a single group model: extinction (Fig. 8.1a), limit-cycle (Fig. 8.1b), and sustainable population (Fig. 8.1c). Similarly, the values of v for scenarios 2 and 3 were selected by observing the proportion of runs where the single group was extinct after 500 time-steps. This helped providing a rough proxy for defining light ($v = 9$, 0.2 extinction rate), intermediate ($v = 16$, 0.5 extinction rate) and severe ($v = 37$, 0.9 extinction rate) disturbance processes (see Fig. 8.1e).

The resulting parameter space (see Table 8.2) has four dimensions and 34 coordinates. For each unique parameter combination, the simulation has been computed 100 times with 500 time steps each. Given that the primary focus of the simulation is to establish the equilibrium properties of the system, the first 200 time-steps have been discarded from the analysis as a "burn-in" stage.

The results of the simulation exercise can be illustrated using a scatter-plot of A_t against A_{t+1}. This data representation can help identify whether the rank-size pattern is stable (point attractor), oscillates between two extremes (limit cycle attractor) or fluctuates chaotically (strange attractor; see McGlade 1995 for a detailed discussion on attractors and their relevance in archaeology), and shows, at the same time, the observed range of variation as well as the frequency and the magnitude of changes (see Fig. 8.2).

---

[1]The parameters defining reproduction (ρ), death (ω_1 and ω_2), cooperation (b), and threshold of evidence (c) can be all aggregated into different types of relationship between key group sizes and net-growth rate. Crema (2013b) showed that the dynamics were significantly different only when the net-growth rate was extremely low and equivalent to zero at the equilibrium group size (i.e. the value of g satisfying the conditions $\phi(g) = \phi(1)$ and $g > 1$). The parameter values chosen for this chapter determines a net growth rate which remains positive above this size.

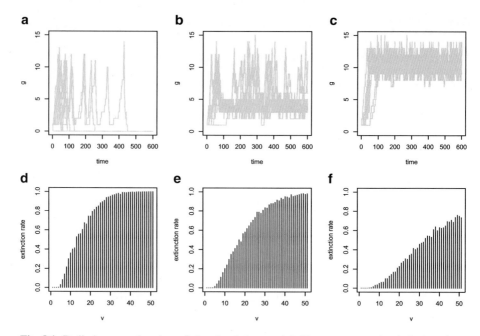

Fig. 8.1 Preliminary exploration of the simulation model. The *upper row* (**a–c**) depicts three sets of ten time-series of population change with the effects of different parameter settings of β ((**a**): $\beta = 0.3$; (**b**): $\beta = 0.35$; (**c**): $\beta = 0.4$). The *lower row* shows the proportion of runs with extinction (among 1,000 simulation runs) for different settings of v and three distinct values of K_{lo} ((**d**): $K_{lo} = 0$; (**e**): $K_{lo} = 10$; (**f**): $K_{lo} = 20$; in all cases K_{hi} was set to 400). In all cases the experiments have been conducted using a single patch world, with the settings listed in Table 8.2

8.4.1 Benchmark Model and Spatial Heterogeneity (Scenarios 0 and 1)

Figure 8.3 shows the parameter space for scenarios 0 and 1. The primary axis of variation in the system behaviour is along an increasing frequency of decision making (z), higher knowledge (k), and wider range of interaction (h), while the effects of increasing heterogeneity of the resource distribution (v) appears to have almost no effect. When z, k, and h have their smallest values, the system is highly disconnected, and the agents distribute themselves to local optima (the best patch around their neighbourhood) leading to the formation of stable convex systems.

The spatial range of interaction plays a pivotal role in this scenario, as increasing values of any of the other three parameters do not affect alone the broad properties of the system (i.e. the type of attractor), expect for larger fluctuations of A around smaller mean values. Once the spatial range of interaction is increased ($h \geq 3$), the implications of the other three parameters become evident in the scatterplots. Agents can now move freely in the landscape and hence the effects of their movement

Table 8.2 List of parameters and values

Symbol	Name	Values
P	Number of patches (cells)	100
μ	Basic fitness	10
b	Benefit of cooperation	0.5
ε	Basic payoff variance	1
ρ	Basic reproductive rate	0.05
ω_1	Death parameter 1	1.2
ω_2	Death parameter 2	5
z	Frequency of decision-making	0.1, 0.5, 1.0
h	Spatial range of interaction	1, 3, 10
k	Sample proportion of observed agents	10^{-8}, 0.5, 1.0
c	Threshold of evidence	3
K	Resource input	200
v_S	Stochastic disturbance parameter (scenario 1)	0, 10, 50
v_T	Stochastic disturbance parameter (scenario 2 and 3)	9, 16, 37
K_{lo}	Lowest possible K	10
K_{hi}	Highest possible K	400
ζ	Intrinsic growth rate of K	2
κ	Carrying capacity of K	200
β	Resource resilience to predation (scenario 4)	0.3, 0.35, 0.4

propagate at larger scales, rather than being absorbed locally. As a consequence of this, we can identify an increase in the possible range of values for A and the occasional appearance of primate systems ($A < 0$). However, in most cases these highly hierarchical settlement systems are unstable, as suggested by the smaller density of points in the lower-left quadrants (see $h \geq 3$, $k \geq 0.5$, $z = 0.5$ in Fig. 8.3).

When the frequency of decision-making is set at its maximum ($z = 1$), the range of spatial interaction is sufficiently high ($h \geq 3$) and the sample proportion of observed agents (k) is equal or larger than 0.5, the system exhibits a limit cycle attractor. The scatterplot also shows how the patterns of these limit cycles are affected by the spatial range of interaction, with $h = 3$ showing more gradual transition between primate and convex pattern, and $h = 100$ characterised by rapid shifts (compare with Fig. 8.2). This cyclical dynamic is derived from the high convergence in the tempo of the decision-making (i.e. all agents move at the same time) and the destination of the migration flow (i.e. all agents move to the same place). Slightly optimal groups are rapidly identified and invaded, triggering a positive feedback, which promotes further migration. This becomes soon unsustainable, and once the destination group becomes too large and fitness starts to decline fission events reset the cycle.

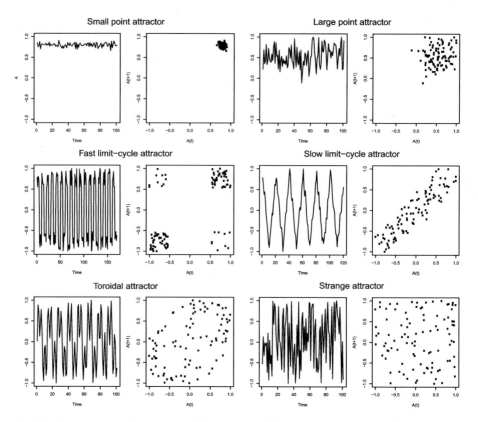

Fig. 8.2 Phase-space scatter plots for different types of time-series (attractors) of A

8.4.2 Temporal and Spatio-Temporal Disturbance (Scenario 2 and Scenario 3)

The pivotal role played by the level of integration between groups is still evident when we add time-varying forms of disturbance processes. Figure 8.4, which depicts the parameter space for scenario 2, shows in fact that low levels of k, z, and h lead to convex point attractors, while their increase determine the emergence of continuous shifts in the rank-size distribution. Details on these shifts are contingent to the time-series of K of individual runs, but we can still identify general trends of regularity (e.g. in $z \geq 0.5$, $k \geq 0.5$, $h \geq 3$), suggested by the higher density of points along the diagonal (see also Fig. 8.2).

When the frequency of decision-making (z) is at its highest value (i.e. the agents respond immediately to the perceived variation in fitness) and the spatial range of interaction (h) is equal or greater than 3, we can observe a limit-cycle attractor with relatively few irregular sudden shifts (lower density of points in the top-left and bottom-right quadrants). However, when $z = 0.5$, the number of unexpected

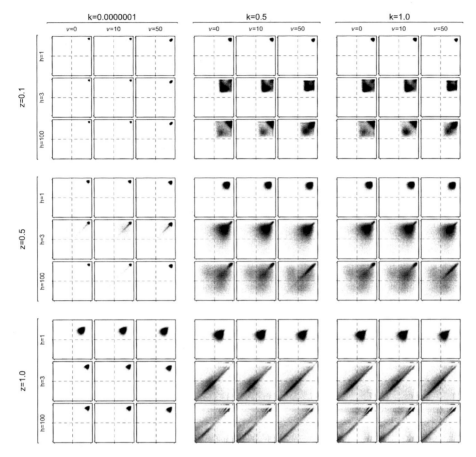

Fig. 8.3 Parameter-space for Scenarios 0 and 1. The x and y axes represent A_t and A_{t+1} and range from -1 to $+1$

changes in the rank-size distribution is much higher, and the system can be classified as a hybrid between limit-cycle and strange attractor. This is most likely explained by a slower response rate of the agents, which are forced to face the consequence of decline (or increase) in K before their relocation. Recall in fact that in the basic, disturbance-free model the system already exhibits high frequency shifts between primate and convex distributions at the highest values of h and z. Hence, within these regions of the parameter space, the disturbance process has a marginal role as the basic dynamics of the system occur at a faster rate. In other words agents relocate themselves before perceiving the consequences of the disturbance events. Conversely, when the response rate is slower ($z = 0.5$), the agents are affected by changes in K. Variations in the abruptness of these disturbance events (v) do not seem to play a significant role other than minor variations in the dispersion of the scatter points: the smallest variation in the availability of resources (K) can be sufficient to induce a cascade effect into the system.

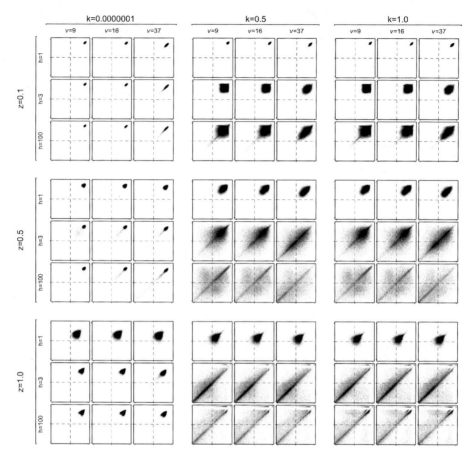

Fig. 8.4 Parameter-space for Scenario 2. The x and y axes represent A_t and A_{t+1} and range from -1 to $+1$

Scenario 3 (Fig. 8.5), which combines both the assumption of spatial hetero-geneity and temporal changes of K, shows similar patterns. Once again the largest variation of the phase-space scatter plot can be observed along the axes defined by h, z, and k. This time, however, increasing values of v exhibit a diagonal "tail" in regions of the parameter space that are characterised by point-attractors in the benchmark model. Observation of individual runs indicates that these pattern are generated from the slow recovery of the system towards highly convex distributions after episodes of sudden decline in A caused by disturbance events. As for scenario 2, the effect of disturbance is tangible mostly for intermediate levels of z, where we can observe increasing episodes of deviations from convex systems with larger values of h and a transition from a "noisy" point attractors to a hybrid between limit cycle and strange attractors. When z is at its highest, the frequency of decision-making is higher than the frequency of disturbance events, leading to a general pattern similar to the benchmark model.

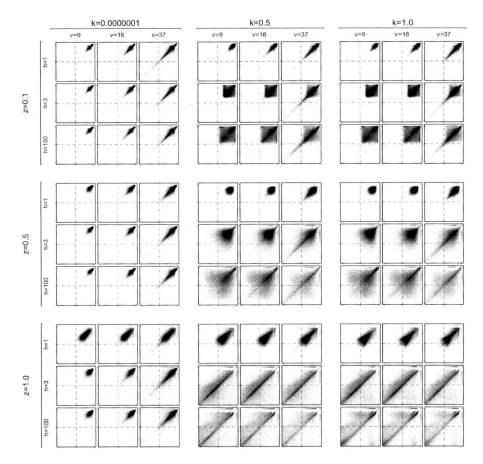

Fig. 8.5 Parameter-space for Scenario 3. The x and y axes represent A_t and A_{t+1} and range from -1 to $+1$

8.4.3   Predator-Prey Model (Scenario 4)

Figure 8.6 illustrates the parameter space for scenario 4, where the amount of resource input K at a given patch is defined by a predator-prey relationship with the group of agents located there. Although this time disturbance can be regarded as endogenous (contra-posed to the exogenous disturbance events of scenarios 1–3), the basic properties of the system remains the same: high levels of h, z, and k still lead to stronger and more frequent variations in the rank size distribution.

The most relevant difference with the other scenarios is how the parameter defining the disturbance process (i.e. the resilience of the resource population β) appears to have a stronger influence in the simulation output. When this is set to the lowest value explored in this series of experiment ($\beta = 0.3$), we observe a larger dispersion of the phase-space scatter plot, often leading to the emergence

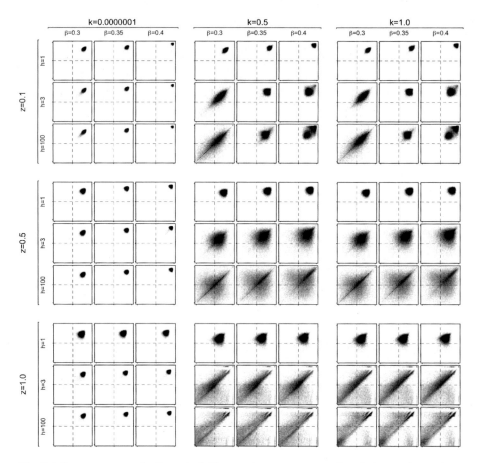

Fig. 8.6 Parameter-space for Scenario 4. The x and y axes represent A_t and A_{t+1} and range from -1 to $+1$

of limit-cycles even when the frequency of decision making is at its lowest. This is the consequence of the ecological inheritance modelled by Eq. (8.7): agents will be subject to a decline in fitness even when there is no internal growth or migration flows. However, if the sample proportion of the observed agents and the spatial range of interaction are also low we still observe exclusively convex point-attractors, confirming once again that the level of integration still plays a pivotal role in explaining the temporal variation in the settlement pattern. When the frequency of decision making is set at its maximum, variations in β do not seem to affect the properties of the system, although the scatter plots appear to be in all cases characterised by a larger number of points outside the main diagonal. As for scenarios 2 and 3 this can be explained by the high response rate of the agents who most likely react to the decline in fitness caused by short-term episodes of overcrowding, rather than actual declines in K.

8.5 Conclusions

This study proposed a model of the emergence and transformation of human settlement pattern by combining a series of assumptions drawn from evolutionary ecology. The two objectives of the simulation exercise were to identify the equilibrium properties of the system in a disturbance-free context (scenario 0), and subsequently explore how four types of perturbations (scenarios 1–4) can affect these. We can summarise the main outcomes of the first objective as follows:

- Convex systems can sustain stable equilibria as long as the level of system connectivity is relatively low;
- Primate systems emerge only temporarily, either as part of a limit-cycle equilibrium, or as a short-term transition from a convex equilibrium. In either case, they require some level of system connectivity, defined here by the spatial range of interaction (h), the frequency of decision-making (z), and the sample proportion of observed neighbour agents (k);
- More generally, increasing connectivity determines an increase in the instability of the system, from a narrowly confined convex point attractor to high frequency oscillations between primate and convex systems, with intermediate states characterised by either point attractors with frequent "escapes" or by more gradual shifts between opposite values of A.

When the system is characterised by low levels of connectivity, individual groups are trapped within local optima while being spatially isolated from each other. Variations in group size will be primarily driven by intrinsic growth rate, as inter-group movement of the agents becomes rare. Interestingly, spatial heterogeneity in the resource distribution (scenario 1) does not affect the dynamics of the system, as agents rapidly find the most suitable locations to settle in the first few runs of the simulation. Once we allow a larger range of interaction, groups become less isolated from each other. This means that small variations in fitness (determined by stochastic components in the model) are amplified by subsequent inter-group migrations. We can conceptualise this with a simple thought experiment. Consider two communities A and B with the same group size. Small differences in the individual yields and random occurrence of reproduction will determine a divergence in their sizes and hence fitness in the short term. Thus, members of A might have a slightly higher fitness than members of B. In the long-term these differences will vanish, but if any individual of B moves to A, this difference will be amplified. Group A will have higher fitness and hence higher chance to further increase its size through reproduction, while members of B will be more attracted by A. If we extend this to a larger number of groups, these dynamics will be further enhanced. As long as the frequency of decision-making is sufficiently high these small variations in the systems will be "detected", and when the spatial range of interaction and the sample proportion of observed agents are both high, chances of agents sharing the same destination becomes increasingly high. This will have a cascade effect, with the sudden appearance of large nucleated settlements (primate systems). These are however highly unstable, and hence their formation will be followed by fission events, which reset and drive the cyclical behaviour of the system.

The low resilience of primate systems is perhaps counter-intuitive, given that in many real-world contexts we can frequently observe these patterns with some degree of stability over time. We should however note that the model proposed here does not allow any form of innovation, and hence the structural properties of the system (defined by the model parameters) remain fixed. This is not the case in many real-world contexts, where the exploitation of novel resources and the adoption of new technology are often enhanced by higher population density (Powell et al. 2009; Lake and Crema 2012). These innovations can easily modify the shape of the fitness curve, allowing for the ability to overcome the problem of declining fitness for larger groups. Furthermore, warfare and direct competitions of resources between different groups can also help in maintaining large nucleated settlements at the expense of others, further allowing the system to prolong primate rank-size patterns. Nevertheless, other studies have demonstrated how these settlements are still destined to collapse or fission in the long-term (Turchin 2003; Griffin 2011), and even when a hierarchical system conserves its shape at the macro-scale, individual communities might be affected by turn-overs, continuously changing their ranks (Batty 2006).

Disturbance processes have a major role when the system is characterised by intermediate levels of connectedness. However, the most relevant conclusion here is that they act as a catalyst rather than being the fundamental cause of shifts in the rank-size distribution. If the necessary preconditions, such as high frequency of decision-making or large spatial range of interaction, do not exist, the system will be almost identical to the expected behaviour in a disturbance-free context. Instead, when these preconditions are met, we can observe a larger number of sudden changes in the rank-size distribution in the form of strange and limit-cycle attractors. However, when the spatial range of interaction, the frequency of decision-making and the sample proportion of observed agents are all low, convex point attractors are minimally affected. Exceptions to this occur only when the abruptness of changes in K is high (e.g. high values of v in scenario 3), but the system will still tend to revert to high values of A in the long term. Similarly, at the opposite end of the parameter space (i.e. when z, h, and k are all high), the benchmark and the disturbance models are almost identical, with only some minor differences observable for scenario 4. This is explained by high rates of settlement reorganisation that, within these regions of the parameter space, would occur regardless of disturbance events.

These conclusions enable us to build a template to which empirical archaeological data can be compared. The abstract nature of the model does not allow us to have precise predictions on the parameters for specific contexts, but, nonetheless, identifying different proxies on the connectivity of the system could provide some clues on why a given rank-sized distribution emerged or changed over time. Thus, one should expect that a rugged landscape might favour the isolation between different communities compared to a plain region, and quantify such an expectation using models of movement based on GIS-led analysis (Conolly and Lake 2006; Bevan 2011) or more complex methods based on circuit-scape theory (McRae et al. 2008). Other proxies for evaluating the degree of system integration include the formal assessment of patterns of cultural similarity or dissimilarity between different communities. Both empirical (Lipo et al. 1997; Shennan and Bentley 2008) and theoretical works (Premo and Scholnick 2011; Crema et al. 2014) have shown great potential for these studies for investigating the strength and variations of regional interaction. Bevan and Wilson (2013) provide a recent example on how these assumptions can also be integrated into realistic models of settlement evolution, allowing for the possibility to directly compare observed data with model predictions.

While these research directions are strongly encouraged, it is also crucial that the underpinning theories of these models are dissected first in artificial and abstract environments where we have full control of each variable. We still need to fully examine the consequences of our theoretical assumptions, before proceeding in applying these models to comprehend real-world changes in human settlement pattern. This chapter is a contribution to such an endeavour.

Acknowledgements I would like to thank the editors for inviting me to contribute to this volume and for providing me rapid feedback during all the editorial process. I am extremely grateful to Andrew Bevan and Mark Lake for discussions and constructive critiques on virtually every aspect of the manuscript and to the two anonymous reviewers for providing valuable comments. This research was funded by a UCL Graduate School Research Scholarship and the AHRC Centre for the Evolution of Cultural Diversity, and acknowledges the use of the UCL *Legion* High Performance Computing Facility (Legion@UCL) and associated support services.

References

Allee WC (1951) The Social Life of Animals. Beacon Press, Boston

Batty M (2006) Rank clocks. Nature 444:592–596

Bevan A (2011) Computational Models for Understanding Movement and Territory. In: Herrera VM, Pérez SC (eds) Tecnologías de Información Geográfica y Análisis Arqueológico del Territorio, Anejos de Archivo Español de Arqueología, Mérida, pp 383–394

Bevan A, Wilson A (2013) Models of settlement hierarchy based on partial evidence. J Archaeol Sci 40(5):2415–2427

Clark CW, Mangel M (1986) The evolutionary advantages of group foraging. Theor Popul Biol 30:45–75

Conolly J, Lake M (2006) Geographical Information Systems in Archaeology. Cambridge University Press, Cambridge

Crema ER (2013a) Cycles of change in Jomon settlement: a case study from eastern Tokyo bay. Antiquity 87(338):1169–1181

Crema ER (2014) A simulation model of fission-fusion dynamics and long-term settlement change. J Archaeol Method Theory 21, 385–404

Crema ER (2013b) Spatial and Temporal Models of Jomon Settlement. Ph.D. thesis, University College London, unpublished

Crema ER, Kerig T, Shennan S (2014) Culture, space, and metapopulation: a simulation-based study for evaluating signals of blending and branching. J Archaeol Sci 43, 289–298

Drennan RD, Peterson CE (2004) Comparing archaeological settlement systems with rank-size graphs: a measure of shape and statistical confidence. J Archaeol Sci 31(5):533–549

Drucker P (1951) The Northern and Central Nootkan Tribes. Smithsonian Institution Bureau of American Ethnology, Washington, Bulletin 144

Endo M (1995) The mobility of resident members of the Ainu in Hokkaido, Japan, in the mid-nineteenth century. Sci Rep Tohoku Univ Ser 7 45(2):75–102

Fagen R (1987) A generalized habitat matching rule. Evol Ecol 1:5–10

Fortin MJ, Dale M (2005) Spatial Analysis: A Guide for Ecologists. Cambridge University Press, Cambridge

Fretwell SD, Lucas HL (1970) On territorial behavior and other factors influencing habitat distribution in birds. I. Theoretical development. Acta Biotheor 19:16–36

Giraldeau LA, Caraco T (2000) Social Foraging Theory. Princeton University Press, Princeton

Gould RA, Yellen JE (1987) Man the hunted: determinants of household spacing in desert and tropical foraging societies. J Anthropol Archaeol 6:77–103

Greene CM, Stamps JA (2001) Habitat selection at low population densities. Ecology 82:2091–2100

Griffin AF (2011) Emergence of fusion/fission cycling and self-organized criticality from a simulation model of early complex polities. J Archaeol Sci 38:873–883

Hawkes K (1992) Sharing and Collective Action. In: Smith EA, Winterhalder B (eds) Evolutionary Ecology and Human Behaviour. Aldine de Gruyter, New York, pp 269–300

Henrich J (2001) Cultural transmission and the diffusion of innovations: adoption dynamics indicate that biased cultural transmission is the predominate force in behavioral change. Am Anthropol 103(4):992–1013

Hodder I (1979) Simulating the Growth of Hierarchies. In: Renfrew C, Cooke KL (eds) Transformtaions: Mathemathical Approaches to Culture Change. Academic, New York, pp 117–144

Johnson GA (1980) Rank-size convexity and system integration: a view from archaeology. Econ Geogr 56:234–247

Jones R (2010) The village and the butterfly: nucleation out of chaos and complexity. Landscapes 1:25–46

Kennett D, Anderson A, Winterhalder B (2006) The Ideal Free Distribution, Food Production, and the Colonization of Oceania. In: Kennett DJ, Winterhalder B (eds) Behavioral Ecology and the Transition to Agriculture. University of California Press, Berkeley, pp 265–288

Kohler TA, Varien MD (2010) A Scale Model of Seven Hundred Years of Farming Settlements in Southwestern Colorado. In: Bandy MS, Fox KR (eds) Becoming Villagers: Comparing Early Village Societies. University of Arizona Press, Tucson, pp 37–61

Lake MW, Crema ER (2012) The cultural evolution of adaptive-trait diversity when resources are uncertain and finite. Adv Complex Syst 15(1 & 2):1150013. DOI 10.1142/S0219525911003323. http://dx.doi.org/10.1142/S0219525911003323

Lipo CP, Madsen ME, Dunnell RC, Hunt T (1997) Population structure, cultural transmission, and frequency seriation. J Anthropol Archaeol 16:301–333

McGlade J (1995) Archaeology and the ecodynamics of human-modified landscapes. Antiquity 69:113–132

McRae BH, Dickson BG, Keitt TH, Shah VB (2008) Using circuit theory to model connectivity in ecology, evolution, and conservation. Ecology 89:2712–2724

Mithen SJ (1990) Thoughtful Foragers: A Study of Prehistoric Decision Making. Cambridge University Press, Cambridge

Pelletier DL, Frongillo EA, Habicht JP (1993) Epidemiologic evidence for a potentiating effect of malnutrition on child mortality. Am J Public Health 83(8):1130–1133

Powell A, Shennan S, Thomas MG (2009) Late Pleistocene demography and the appearance of modern human behavior. Science 324:1298–1301

Premo LS (2010) Equifinality and Explanation: The Role of Agent-Based Modeling in Postpositivist Archaeology. In: Costopoulos A, Lake M (eds) Simulating Change: Archaeology into the Twenty-First Century. University of Utah Press, Salt Lake City, pp 28–37

Premo LS, Scholnick JB (2011) The spatial scale of social learning affects cultural diversity. Am Antiq 76(1):163–176. http://saa.metapress.com/index/A661T246K0J1227K.pdf

R Core Team (2013) R: A Language and Environment for Statistical Computing. R Foundation for Statistical Computing. Software. http://www.R-project.org

Roberts BK (1996) Landscapes of Settlement: Prehistory to the Present. Routledge, London

Savage SH (1997) Assessing departures from log-normality in the rank-size rule. J Archaeol Sci 24:233–244

Shennan SJ, Bentley AM (2008) Interaction, and Demography Among the Earliest Farmers of Central Europe. In: O'Brien MJ (ed) Cultural Transmission and Archaeology: Issues and Case Studies. SAA Press, Washington, pp 164–177

Sibly RM (1983) Optimal group size is unstable. Anim Behav 31(3):947–948

Sutherland WJ (1983) Aggregation and the "ideal free" distribution. J Anim Ecol 52:821–828

Tregenza T (1995) Building on the ideal free distribution. Adv Ecol Res 26:253–302

Turchin P (2003) Historical Dynamics: Why States Rise and Fall. Princeton University Press, Princeton

Verhulst PF (1838) Notice sur la loi que la population poursuit dans son accroissement. Correspondance Math. Phys. 10:112–121

Watanabe H (1986) Community Habitation and Food Gathering in Prehistoric Japan: An Ethnographic Interpretation of the Archaeological Evidence. In: Pearson RJ, Barnes GL, Hutterer KL (eds) Windows on the Japanese Past: Studies in Archaeology and Prehistory. Centre for Japanese Studies, University of Michigan, Ann Arbor, pp 229–254

Winterhalder B, Kennett DJ, Grote MN, Bartruff J (2010) Ideal free settlement of California's Northern Channel Islands. J Anthropol Archaeol 29:469–490

Zipf GK (1949) Human Behavior and the Principle of Least Effort. Harvard University Press, Cambridge

Chapter 9
Understanding the Iron Age Economy: Sustainability of Agricultural Practices under Stable Population Growth

Alžběta Danielisová, Kamila Olševičová, Richard Cimler, and Tomáš Machálek

9.1 Introduction

When searching for the explanation of subsistence strategies in different social groups of a complex society, the key factor is the relationship between its agricultural base and the social hierarchy of settlements. Due to the fragmented nature of data available such presumptions were mostly only theoretical and often failed to effectively capture the complexity of the system under consideration. As a consequence, we have only a limited picture of how societies may have functioned in past. In order to capture the whole complexity of subsistence, more sophisticated methods and tools are needed.

Recent studies show that, on top of a comprehensive collection of data, the building of explanatory models is a valid way of exploring the complexity of past societies (e.g. Kohler and van der Leeuw 2007). Models are valid scientific tools, which not only can help to understand human-landscape interactions and processes of change, but are also very powerful for suggesting directions for future research (Wainwright 2008), by challenging traditional ideas and exploring new issues (Wright 2007, p. 231). At the same time, models do not attempt to become exact reflections of the past reality; they just aim to provide more precise research questions, select appropriate attributes and factors to be further examined on an accurate spatial and temporal scale. This framework can be used to investigate socio-ecological interactions over a broad range of social, spatial and temporal

A. Danielisová (✉)
Institute of Archaeology of the Czech Academy of Sciences, Prague, Czech Republic
e-mail: danielisova@arup.cas.cz

K. Olševičová • R. Cimler • T. Machálek
University of Hradec Králové, Hradec Králové, Czech Republic
e-mail: kamila.olsevicova@uhk.cz; richard.cimler@uhk.cz; tomas.machalek@gmail.com

© Springer International Publishing Switzerland 2015   183
G. Wurzer et al. (eds.), *Agent-based Modeling and Simulation in Archaeology*,
Advances in Geographic Information Science, DOI 10.1007/978-3-319-00008-4_9

scales, allowing for a wide range of past archaeological issues to be addressed. Agent-based models (ABM) provide an invaluable way to explore and test various theoretical hypotheses, including cases where detailed data is not available from the archaeological record.

This chapter deals with the subsistence strategies and the economic background of late Iron Age oppida in central Europe, the fortified agglomerations occupied in the last two centuries BC. Oppida settlements represent complex systems (societies) with multiple functions. They appeared as a part of an economically advanced environment, together with a distinctive intensification of settlement patterns. However, the oppida were mostly built in landscapes considered to have been marginal in regard to what appeared to be the common settlement strategy. Due to the specific location and the widespread evidence of late La Tène open lowland settlements (which were believed to supply the oppida with the necessary food resources) the agricultural potential of the oppida has usually been challenged (e.g. Salač 2000, 2006). According to the archaeological record, their settlement density in the late La Tène period increased over a short time span and then decreased again. This probably quite rapid depopulation and collapse of these agglomerations during the second half of first century BC was attributed, among all, to a supply crisis. In short: the oppida agglomerations were perceived as too "specialized" and therefore engaged in other activities than agriculture. Because of that, they were not capable of producing sufficient food. This fact should have eventually contributed decisively to the collapse of the La Tène social structure as a whole. Interestingly though, the archaeological record reveals clear signs of some engagement in agricultural production—by the evidence of crops cultivation, livestock breeding or traditional agricultural household units (cf. Küster 1993; Danielisová and Hajnalová in print). Therefore, the idea of the oppida acting as pure receivers of the agricultural products needs reconsideration. Causes for a gradual depopulation trend of the fortified settlements can be attributed to several factors both endogenous and exogenous. However, beside political (reaction to the military events), economic/commercial (difficulties on long distance commercial routes) and organizational (less people to perform necessary tasks), the ecological/subsistence problems are worth contemplation as well.

The relationship between population growth and the development of the society depends on the availability of basic resources from the environment; however, this relationship is in fact never constant. In the models of social complexity, which include and interconnect innovation, specialisation, political structure, market integration, but also migration, changes in settlement patterns and abandonment of settlements, population growth and over-exploitation of natural resources play an important role from which wide range of social phenomena have been explained (Bayliss-Smith 1978, p. 130). According to historical sources, exceeding the appropriate carrying capacity was not a rare occasion in history (cf. Schreg 2011, p. 312) even in societies with developed market networks. Intensification of production led to innovations in agriculture on the one hand, but also to a more rapid depletion of the land resources, especially where their extent was limited, on the other hand. This prompted behaviour which could have led to more profound social change at the end of the Iron Age in central Europe. Our main objective is to approach this issue by the modelling of population dynamics, and subsequently by the modelling of agricultural strategies and socio-economic interactions. This chapter presents the initial models focused at the moment on the oppidum's own agricultural production, i.e. a society pursuing agro-pastoral activities within the given temporal and spatial scale which is tested against subsistence, surplus production and carrying capacity factors. We aim to find the limit of the environment for the growing population at a certain point of time. Some of the fundamental questions being asked at that stage include:

1. Using what cultivation strategies can the population most effectively exploit natural resources in order to be self-sufficient?
2. What are the dynamics of production with constantly growing population (subsistence—surplus—success rate—diminishing returns)?
3. What is the maximum population that can be sustained in a given environment and when was this maximum reached?

In order to credibly recreate the economic life of an agglomeration in its specific environment, a coupled environmental and social modelling approach is employed. In more detail, we incorporate a GIS model representing the environmental conditions and an agent-based model addressing social issues (the population dynamics) and land-use patterns. As outcome, these models can provide substantial information about the limits of rural economy at a particular location and they may help to determine possible stress situations. As result we may be able to explain the reason for population decline, in case it was caused by reaching the limits of food production.

9.2 Methodological Approach

9.2.1 Theoretical Framework

The relationship between population growth and development of the society depends on the availability of basic resources from a given unit of land, the ability of the population to exploit them as well as to socially organise this process. In this chapter, some models of interactions between the settlement and its natural environment are applied, including "settlement ecosystems" (the systems of land use based on the relationship between the landscape and human decision how to exploit it, cf. Ebersbach 2002), "population pressure" (the point in population growth when it is larger than the resource base, e.g. Bayliss-Smith 1978) and "carrying capacity" (the limit above which the population must either diminish or innovate their exploitation strategies, cf. Del Monte-Luna et al. 2004). All of these concepts will be tested against archaeological and environmental data. As precondition, we must define resource levels of the ecosystems (i.e. productivity of the land), productivity potential of the population exploiting these resources (i.e. labour input, technology and task management) and test if and under what conditions certain resources can become limiting factors, and what implications can be derived from that (e.g. adoption of new subsistence strategies, new technologies, commercial contacts, social transformations, settlement abandonment etc.).

By modelling subsistence strategies, we seek to explore the following questions:

- the size of the community and proportion of people engaged in agricultural work
- the organisation of the working process
- demands concerning land and labour
- the carrying capacity threshold
- the scale of sustainable agricultural production
- the stability of production (number of stress situations)

The theoretical framework for modelling the oppidum's own agricultural production encompasses the following tasks:

1. Modelling the population and its subsistence needs:

   a. Estimating the population size and its dynamics due to demographic processes.
   b. Estimating the available work force in relation to the population dynamics.
   c. Calculating the annual nutrient demands of the population.

2. Modelling crucial resources:

   a. Deliminating the maximum distance people would have been willing to go in order to cover their subsistence needs.
   b. Predicting the hinterland components (fields, fallows, pastures, woodlands etc.) according to the environmental conditions (a non-continuous pattern of arable land use zones).
   c. Organization of the working process in relation to the land-use patterns.
   d. Within the predicted hinterland, there should not be any competing sites (considered as contemporary), otherwise we would have to: (1) enlarge the settlements' catchment area, (2) consider a cooperative/competitive relationship between multiple sites.

3. Modelling resource exploitation and assessing its limits:

   a. Estimating the energy potential of key resources (yields of fields, pastures, fodder, woodlands ...).
   b. Outlining possible exploitation and production strategies.
   c. Comparing the number of consumers and their needs in relation to the production.
   d. Relating the labour input to the production.
   e. Determining possible insufficient or lacking resources or productivity (labour input) and the impact of that factor.

4. Modelling the dynamics of the production:

   a. Including the fluctuations of the harvests according to different farming strategies.
   b. Measuring the actual levels of harvest, surplus and potential storage reserves.
   c. Determining the agricultural sustainability (number of stress situations).
   d. Detecting the limit threshold, estimating the carrying capacity of the hinterland in relation to the population pressure and the potential scarcity or depletion of resources.
   e. Count in the external factors, if there were any (climate variations, conflicts, socio-political decisions).

9.2.2 Modelling Tools

In context of our agent-based modelling approach, agents may represent individuals, families and collectives, management agencies, and policy making bodies, all of whom are able to make decisions or take actions that affect development of the society and changes the environment. Because agent-based models are time-based, the development of the society can be studied for over 100 years in course of its evolution. Components of models include ecological, socio-economic and politico-cultural parameters addressing individual questions, which can be developed and implemented gradually by producing models that target individual questions separately.

Models are implemented in the agent-based modelling software package Netlogo (Wilensky 1999), its plug-ins and extensions:

- *BehaviorSpace* enables repeated run of simulations with different settings of parameters. Output data can be stored and further processed using statistical toolkits.
- *BehaviorSearch* (Stonedahl 2011) is a tool which allows automated search for (near) optimal values of parameters of models with respect to objective function. The search process is performed with evolution computation techniques, which are widely used for optimization problems. The nature of the algorithm does not guarantee that the calculation always converges to the global optimum, but on the other side, it is general enough to be able to solve wide scale of problems. For demonstration of BehaviorSearch, see the replication of the "Artificial Anasazi" Project (Stonedahl and Wilensky 2010).
- *Fuzzy-plug-in* (Machálek et al. 2013) provides fuzzy-related functionality. It is based on a FuzzyLogic library by Cingolani and Alcalá-Fdez (2012). The plug-in was implemented for the purpose of our research.

The advantage of the simulation process lies in a relatively fast transcription of parameters into the programming language. Many different scenarios can be created by alteration, adaptation and combination of the input data. Therefore, we can aspire to simulate quite complex phenomena (such as the "agricultural year"). A more difficult task is the definition of relevant research questions, isolation of valid parameters and the actual definition, execution, evaluation and recurrent adaptation of the created models (also see Chap. 4). Outputs can be compared to a broad framework of data established by combining the results from archaeological excavations, archaeological surveys and regional-scale environmental studies.

9.3   Data Resources and Modelling Inputs

The models are based on the region around the oppidum of Staré Hradisko (Czech Republic). It offers quite complex archaeological and environmental data and analyses carried out upon the material collected during the long-term excavations (cf. Čižmář 2005; Meduna 1961, 1970a,b; Danielisová 2006; Danielisová and Hajnalová in print). Sections 9.3.1–9.3.4 summarise the sources of the environmental input variables and agricultural production processes used in our study.

9.3.1   Environmental Data and Ecology

Given all the required input parameters for the modelling of the landscape exploitation, the most relevant data that reflected the oppidum's setting needed to be obtained (cf. Table 9.1). Because detailed LIDAR scans were not available for this location, a digital elevation model (DEM) of the relief was computed using local

Table 9.1   Environmental input variables

Environmental variables	Source
Relief (digital elevation model—DEM)	Modelled in GIS from 1:5,000 topographical maps (ArcGIS, resolution 5 × 5 m)
Landforms, topographical features, topographic wetness index	Modelled in GIS from DEM (ArcGIS, IDRISI, Whitebox, Landserf)
Hydrology	Modelled in GIS from DEM and complemented by fluvial sediments in geological maps and historic mapping (ArcGIS, manual correction)
Geology and soils	Digitised from geological and soil maps 1:50,000
Soil quality	BPEJ soil evaluation
Potential vegetation	From Neuhäuslová (2001)
Climate	Macrophysical Climate Model (MCM) created from local meteorological data, cf. Bryson and DeWall (2007), Danielisová and Hajnalová (in print)
Weather	Recorded historic frequency of the "events" (hailstorms, late frosts, heavy rains …), from Brázdil et al. (2006)

contour maps. The DEM data were used to form secondary variables such as the terrain settings, the location of landforms (slope gradient, aspect, local elevation difference), and the topographical features. The hydrologic settings were derived from DEM and completed by the geologic and historic mapping in order to recreate (as closely as possible) the original stream network irretrievably altered by current agronomic practices (e.g. melioration). Location of streams can help to predict the location of settlements and the activities requiring the proximity of water source (e.g. pastures). Hydrologic modelling was used as well to the computation of the topographic wetness index. This variable describes the propensity for a land plot to be saturated by the runoff water, given its flow accumulation area and local slope characteristics (Lindsay 2012). It is used to locate the areas (from the topographical point of view classified simply as flat) with tendency to be wet or dry. Soil coverage, geology and the potential vegetation should also contribute, as input data, to addressing the agricultural potential of an area.

It should be noted that these variables are interconnected. Since one is often derived from the other, their impact can multiply in GIS modelling. Geology was therefore used to delimit the individual structures (alluvial plains, rocks etc.) and potential vegetation served as secondary evidence to the soil data. Suitability of the soils for farming is measured by the soil type (light Cambisols, Gleys or Pseudogleys, colluvial deposits and fluvial deposits in the floodplains), the amount of humus, their depth and rockiness, accessible through the so-called BPEJ evaluation of the farmland. Modelled local MCM (Macrophysical Climate Model) mean annual temperatures and annual precipitation for the oppidum of Staré Hradisko showed warmer and to a certain extent also wetter conditions during the occupation of the oppidum in comparison to modern times (Danielisová and Hajnalová in print). This model also addresses the potential evapotranspiration, calculated from the annual temperature and precipitation. It reflects the general tendency of the climate to be "more wet" (colder, wetter) or "more dry" (warmer, less precipitation) which can influence the location of the farming plots (e.g. avoiding the wet plains etc.).

9.3.2 Socio-Economic Variables

The occupational span of the oppidum is around 150 BC to 50–30 BC, i.e. ca. 100–120 years of existence. According to the chronologically significant material (cf. Table 9.2), the population density was increasing during the initial 2–3 generations (50–80 years), until it reached the point when it started to decrease again (ca. 80–70 BC). This decline seems to have been quite rapid (1–2 generations); the final population might have been even five times smaller than during its highest density. Concerning exact numbers, unfortunately, only approximate data is available, but given the archaeological evidence (number of households, chronologically significant artefacts) the initial population density should have certainly comprised several hundreds of people.

Table 9.2 Input variables and information on the anthropogenic aspects of the models (for the definition of "strongforce" and "weakforce" see Sect. 9.4.1)

Population and economical structures

Social/economic variable	Source
Population density	Settlement structure vs. chronology of occupation (personal items—e.g. brooches, frequency of coins, amounts of refuse in settlement stratigraphy, number of households)
Cultivated crops	Archaeobotanical collection (Hajnalová 2003)
Storage devices	Settlement structure (Danielisová 2006)
Stock keeping	Archaeozoological collection (Peške, unpublished data)
Agricultural tools (ploughshares, hoes, harrows, scythes, leaf foddering knives, rotary querns)	Material collections (Meduna 1961, 1970a,b; Čižmář 2005; Danielisová and Hajnalová in print)
Distance costs and hinterland area	Walking distance from the settlement (Chisholm 1979), algorithm (Gorenflo and Gale 1990)
Settlement ecosystems	Agricultural strategy in relation to the environment and the production goals (Ebersbach 2002; Schreg 2011)

Agricultural practices

Activity, practice	Source
Intensity of land-use (agricultural strategies)	Intensive-extensive-mixed (Bakels 2009; Halstead 1995; Danielisová and Hajnalová in print)
Field plot sizes	0.235–0.94 ha/person including seed corn and losses (Halstead 1987, 1995; Dreslerová 1995)
Fallowing	Fields: fallows = 1:3 (Fischer et al. 2010; Halstead and Jones 1989)
Crop yields	700–3,000 kg/ha (Bogaard 2004; Hejcman and Kunzová 2010; Kunzová and Hejcman 2009; Rothamstead Research 2006)
Ploughing rates	0.21–0.42 "strongforce"*/ha (Halstead 1995; Russell 1988)
Hoeing, harrowing, weeding rates	0.85–1.27 "weakforce"*/ha (Halstead 1995; Russell 1988)
Manuring	(Minimum) N = 46 kg/ha (Hejcman and Kunzová 2010; Kunzová and Hejcman 2009; Bogaard et al. 2006; Fischer et al. 2010)
Harvest and crop processing rates	0.42 "strongforce" + 0.85 "weakforce"/ha (Halstead 1995; Halstead and Jones 1989; Russell 1988)

The increasing population trend should be reflected in models of food and fodder production (i.e. the spatial change in the field, pasture, and forest area) as well as in the numbers of livestock.[1] Data derived from the archaeological record are used to create anthropogenic aspects of the models and form model inputs. Ethnographic data can clarify agricultural strategies and processes as well as add new details. Table 9.2 shows an overview of the input variables and their sources.

---

[1] Livestock-woodland models (currently under development) are not presented in this chapter.

Potential evidence of agricultural activities carried out by the oppidum inhabitants can be indicated by particular material groups, archaeobotanical and archaeozoological assemblages, and settlement features (especially storage facilities), (cf. Danielisová 2006; Danielisová and Hajnalová in print). From the cultivated plants, the majority in the archaeobotanical assemblages is represented by glume species (barley—54 %, and spelt—24 %) and to a lesser extent by free-threshing wheat (13 %). Pulses have also been attested. Enclosed farmsteads at the oppidum revealed clear evidence of surface storage devices (granaries). There is neither the evidence of sunken silos, which would enable long-term storage of cereals, nor communal storage facilities, which would indicate bulk supplying of the whole community. Concerning the provisioning for the individual households, it can be said that they were likely in charge themselves.

9.3.3   The Agricultural Hinterland

All societies in the past aimed at obtaining their necessary natural resources from the immediate vicinity. The method of Site Catchment Analysis (Higgs and Vita-Finzi 1972) was used for modelling the oppidum's hinterland (Fig. 9.1). This approach is based on models of economics and ecological energy expenditure, and provides a framework within which the economic activities of a particular site can be related to the resource potential of the surrounding area. We thus needed to delimit the "easily accessible" area in the site's surroundings, which would have encompassed fields/fallows, pastures, meadows and managed forests. Considering the locational rules of the "least effort models" and the variable topography, the area was modelled as cost distance according to walking speed from the centre (Fig. 9.1a). A formula of travelling velocity V across terrain suggested by Gorenflo and Gale (1990, p. 244) was used to delimit one hour of walking distance:

$$V = 6 \cdot e^{-3.5([slope\ in\ \%]+0.05)}$$

This roughly corresponds to a distance of 4–5 km generally considered as a threshold for the travelling on a daily or semi-daily return (cf. Chisholm 1979, p. 64).

The criteria for the prediction of fields were related to the environmental variables: topography, soils, and climate (cf. Table 9.1 and Sect. 9.3.1). The fields had to be placed on fairly moderate slopes—less than 5°, 5°–10° and 10°–15°[2] respectively. Together with the other variables "aspect", "soils" (quality, depth,

---

[2]Opinions on how steep slopes are still cultivable quite differ. In the agricultural models it was believed that slopes beyond 7°–10° were not tillable, though this applies especially for the machinery not the manual cultivation (cf. Fischer et al. 2010). These (indeed arbitrary) categories deliberately represent more benevolent option suited for hand tools (<15°) and ploughing animals (<10°).

rockiness), "topography", and "wetness"[3] it was put together through the Multi-Criteria Evaluation analysis (Eastman 2006, pp. 126–134) by which different field suitability categories were created (Fig. 9.1b). The plots classified as unsuitable (too wet, too rocky or on slopes too steep) were excluded from the field model. One of the crucial factors for the prediction of fields was the accessibility from the settlement. Therefore most suitable areas were plotted as the most fertile zones located as close as possible to the settlement. A cost penalty was included in the agricultural model for fields exceeding the distance of 2 km (cf. Chisholm 1979, p. 61). This option applies especially for more intensive regimes of land-use; the cost impact was lower for fields where extensive practices were carried out. Fields within distant zones could have been subjected to different land-use and management types—more intensive closer to the oppidum and more extensive further away (cf. Sect. 9.3.4). The remaining terrain can be attributed to open and forest pastures, forest openings and woodlands.

9.3.4 Production Processes

Labour rates and resources required for each of the agricultural tasks listed in Table 9.2 have been studied (Russell 1988; Halstead 1995; Halstead and Jones 1989 etc.) and can be used for modelling of farming practices. The default presumption for the model is that households used animal traction for cultivating their fields. The actual area of fields, as well as the labour input per unit area, varies greatly according to the number of inhabitants and different arable farming strategies employed. With higher yields during an increasing intensity of cultivation, the area of fields could have decreased and vice versa. High annual harvest fluctuations are apparent in modern agricultural experiments (e.g. Rothamsted Research 2006; Hejcman and Kunzová 2010; Kunzová and Hejcman 2009). Variable annual yields are also being regularly mentioned in the historic records (cf. Campbell 2007; Erdkamp 2005). Therefore, using mean yield estimation in archaeological modelling would provide only a static indication of production. A relative structure of inter-annual fluctuations in the ancient crop yields from a particular area may be established by extrapolating from modern or historical data, preferably from the same region and without estimating any absolute mean value (Halstead and O'Shea 1989, p. 6; Hejcman and Kunzová 2010). A general range between 500 and 3,000 kg/ha (Danielisová and Hajnalová in print) can be considered

---

[3]The categorization (with decreasing "suitability") of the variables was the following: Aspect: slopes exposed to the cardinal points (from the South to the North); soil quality: Cambisols, Cambisols with Gley, Gleys and Pseudogleys, fluvial deposits in the floodplains; soil depth and rockiness: low, moderate, high (according to the BPEJ land evaluation); topographic wetness index: delimited the areas which were "too wet" (especially in the floodplains; more detailed categorization of this variable will be useful especially for the modeling of the pastures); topography: excluded slopes too steep (>15°) or areas too rocky etc.

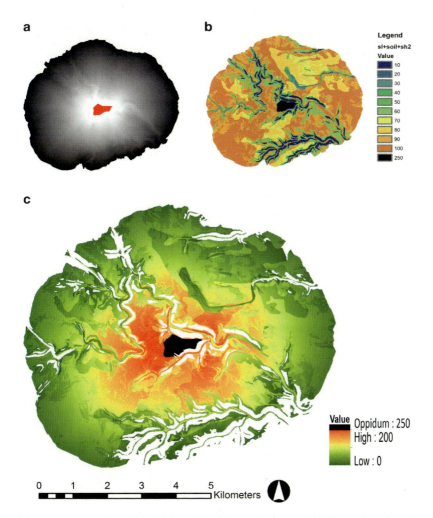

Fig. 9.1 Quantitative GIS model of the oppidum's hinterland area within its predicted catchment: (**a**) cost distance from the settlement, (**b**) evaluation of land suitability for the cultivation of crops, (**c**) resulting model of the hinterland's suitability for agricultural tasks. The "utility index" ranges from 0 to 250 (0–200 represents increasing suitability, 250 is the oppidum)

a suitable variance of general yield variability, derived from the information on local environmental and climate conditions, the reconstructed scale and intensity of farming (by "intensity", we understand the amount of labour input required to process one unit area of land) and production targets (from small subsistence needs to surplus production requirements). The following three agricultural strategies are assumed to have been possibly practised by the Iron Age population (also see Table 9.2 for a reference regarding input values):

1. Intensive farming on small plots: fields were manured by grazing and stable dung; they were intensively tilled by hand, and weeded. Working animals could be used for maintaining higher yields; also, rotation of crops (cereals, pulses) was practised. An intensive farming strategy represents the most labour-demanding option, which tends to be limited in scale or covers only the subsistence needs. Larger production (in the meaning of surplus production) would require higher labour input at the expense of the other activities (like stock farming).

2. Extensive farming on large plots: fields included fallows and were managed less intensively. They were manured especially by grazing animals. The plots could be usually under continuous cropping (i.e. no crops rotation) as the periods of fallow allowed for the sufficient regeneration. An extensive strategy could have been employed especially when the available land was abundant, population pressure low, labour was engaged elsewhere, or it was more preferred than the intensive production. With this strategy the potential for surplus production was higher, but could fluctuate heavily.

3. Mixed strategy: this comprised a combined approach of land managed more intensively within the infields (closer to the settlement) and more extensively in the outfields (further from the settlement). Both were ploughed by working animals. Infields would be fertilised by the farmyard manure, could be weeded or hand tilled. Crop there was rotated as under the intensive strategy. For the fields further away, fallows were included (management of which was less intensive).

The extensive strategy, as well as the mixed one, required quite large areas to be available without any competing sites around. When population growth caused pressure on resources, extensively managed fields further away could be turned into more intensively managed ones. If the increasing intensity of land cultivation could not be matched by either adequate labour input or numbers of animals (to secure necessary manuring), a stress situation would develop.

9.4 Models

In the models of social complexity, population growth plays an important role. Population pressure and over-exploitation of resources are very important concepts from which a wide range of social phenomena can be explained (Bayliss-Smith 1978, p. 130). Our primary objective int that context was to find a stable and reliable model of the population growth, with matching initial and final age distributions. The simulation of the population dynamics (Sect. 9.4.1) provided input data for further modelling and investigation of the oppidum's production potential and carrying capacity threshold of the hinterland (Sects. 9.4.2 and 9.4.3). Agricultural production around the oppidum is perceived as having developed from the beginning of land cultivation, relative to the beginning of the occupation of the oppidum, and having lasted for about 100–120 years.

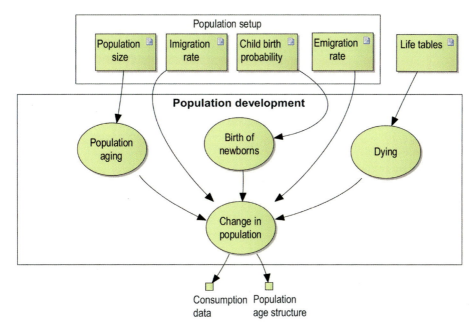

Fig. 9.2 Population model

9.4.1   Population Dynamics Model

Population growth is defined as the change in number of individuals over time (Fig. 9.2). It is assumed that all populations grow (or decline) exponentially (or logarithmically), unless affected by other forces. The simplest Malthusian growth model assumes the exponential growth is

$$P(t) = P_0 e^{rt}$$

where P_0 is the initial population, r is the growth rate and t is time. In our case, the initial number of inhabitants is said to be between 500 and 800, the maximum number of inhabitants after 100–120 years is between 2,000 and 5,000. The rapid annual growth can in suitable circumstances reach 2 % (cf. Turchin 2009, p. 12).

Our model has one type of agent representing inhabitants. Each such "inhabitant-agent" is characterized by gender (male, female), age (discrete value), and age-category (suckling, toddler, child, older child, young adult, adult or elder). Auxiliary variables were added for monitoring characteristics of the whole population: percent-of-suckling, percent-of-toddlers, percent-of-children, percent-of-older-children, percent-of-young-adults, percent-of-adults, percent-of-elders. Summarizing variables num-of-inhabitants, actual-workforce and actual-consumption inform about the structure the population.

The model has the following input parameters:

Table 9.3 Proportional distributions of the age groups

Age group	% in population
Suckling (0–1)	11.43
Toddlers (1–3)	5.71
Small children (3–10)	14.29
Older children (10–14)	11.43
Young adults (15–19)	8.57
Adults (19–48)	37.14
Elderly (48–)	11.43

Table 9.4 Daily caloric requirement for different sex and age groups (after Gregg 1988, Table 1, modified)

Age group	kcal/day
Suckling (0–1)	0
Toddlers (1–3)	1,360
Small children (3–10)	2,000
Older children (10–14)	2,300–2,500
Young adults (15–19)	2,500–3,000
Female adults (19–48)	2,600
Male adults (19–48)	3,000
Elderly (over 49)	2,000

- initial population size between 500 and 800 individuals,
- initial population age structure,
- abridged life tables interpolated to a full variant,
- probability Q of a woman having a child in a specific year.

The initial population consist of seven age-groups (suckling, toddlers, children, older children, young adults, adults and elderly), the proportional distribution of each age-group was defined experimentally (see Table 9.3) (Olševičová et al. 2012).

Depending on each sex/age category, one person should yearly consume his/her *required amount of cereals* × *their caloric value* (1 kg of wheat = ca. 3440 kcal).[4] The cereals are assumed to cover 70–75 % of daily energy intake; the rest was supplemented by proteins and other nutrients (cf. Table 9.4).

During the initialization of the model, a population of the supplied size is created. Each inhabitant-agent is assigned an age and gender; globally, counters regarding the number of inhabitant-agents in each age and gender group are updated. The main procedure simulates 120 time steps (see Listing 9.1). At each time step, each inhabitant-agent applies its *get-older* procedure. The procedure operates with abridged life-tables, adopted from the regional model life-tables created by Coale and Demeny for the ancient Roman population (Saller 1994). We used the Model Life Tables Level 3 a 6 West. To complete missing values in the tables (as they were in 5-year intervals), the Elandt-Johnson estimation method (Baili et al. 2005) was

---

[4]For example in case a male adult agent requires 3,000 kcal daily, this means 1,095,000 kcal yearly, which equals 322.05 kg of cereals/year (if only cereals are consumed).

applied. The inhabitant-agent, representing women between 15 and 49 years, also executes the *birth-rate* procedure (avg. 5.1 children per woman) that operates with probability Q.

The model ignores more detailed aspects like partner selection and proportions of various families' formation (nuclear-extended), as those are the variables which would have to be set arbitrarily, without sufficient supporting data. While population size, life tables and population structure were available to us, it was very complicated to estimate the probability Q without relevant statistical data. It should be noted that the probability Q is in fact composed of multiple components, some of them depending on others. Biological fertility of both partners can be estimated (in general, it is decreasing with age), but such a parameter itself cannot explain the whole Q, as there are many socio-economic and other biological factors. As a consequence, NetLogo's BehaviorSearch tool was used to identify parameters of Q experimentally, using genetic algorithms.

Listing 9.1 Population model pseudocode

```
model-setup
  load life tables
  create initial population of N inhabitant-agents
    for each inhabitant-agent
      set initial age
      set gender (male or female)
  set global variables
    count number of suckling
    count number of toddlers
    count number of children
    count number of older children
    count number of young adults
    count number of adults
    count number of elders
  update plots and monitors
end

model-go
  repeat for 120 steps
    set global variables
      count number of weak workforce
      count number of strong workforce
      count total consumption
    for each inhabitant-agent
      get-older
    for each female-inhabitant-agent
      birth-rate
    set global variables
      count number of suckling
      count number of toddlers
      count number of children
      count number of older children
      count number of young adults
      count number of adults
      count number of elders
```

```
    update plots and monitors
  export population data
  export consumption data
  export workforce data
end
```

The probability Q is a function of a woman's age, with the additional limitation that it is defined to be non-zero only in the interval 15 to 49 years.

Based on empirical findings (fertility rate around 5.1; more than two children rarely survived infancy), we have decided to discretize the function Q using intervals of 5 years. This means that probability q_i applies for ages $i, \ldots, i+4$. Taking into account the fact that the time step of the model is one year and that the interval where Q is non-zero from age 15 to age 49, we actually look at a vector of size 7. Individual probabilities are discretized, too—we consider only integer values $\{0,1,2,\ldots,100\}$. Based on that, we can define following parameter vector:

$$Q = (q_{15}, q_{20}, q_{25}, \ldots, q_{45}), \ q_i \in \{0, 1, 2, \ldots, 100\} \tag{9.1}$$

To be able to use BehaviorSearch, an objective function must be also defined. Our objective was to optimize parameters such that a defined population size was achievable; as constraint, the population structure should be as close as possible to the original one. Hence, each of these two aspects were modelled as an own objective function. Function F_1 given in Eq. (9.2) is the first one, giving the percentage of change in final population x considering an initial population A.

$$F_1 = \frac{|x - A|}{A} \tag{9.2}$$

The second aspect is covered in function F_1 (Eq. (9.3)), which represents the average percentage change in seven population's age intervals (suckling, toddlers, children, older children, young adults, adults, elders). More specifically, the change in percentage within the individual age and sex groups was compared to the initial population structure.

$$F_2 = \frac{1}{7} \sum_{i=1}^{7} \frac{|x_i - x_i'|}{x_i'} \tag{9.3}$$

However, BehaviorSearch cannot optimize multi-objective functions; therefore the problem must be converted into a single-objective optimization. We have used simple Euclidean distance:

$$F = \sqrt{F_1^2 + F_2^2} \tag{9.4}$$

In the overall objective function F (Eq. (9.4)), both contained objective functions have the same weight. It should be also noted that the function F can no longer be interpreted as a percentage, as opposed to functions F_1 and F_2.

The model has the two following outputs:

- necessary energy input (caloric input value extrapolated from the actual oppidum population in all sex/age groups) for each year of simulation (Fig. 9.3a).
- available workforce (actual number of people in productive age in particular age/sex categories) for each year of the simulation (Fig. 9.3b). Two main categories were distinguished: *"strongforce"* (males and young males who can perform heavier task such as ploughing, harvesting with scythe, trees cutting etc.) and *"weakforce"* (other age/sex categories, except very small children, who can pursue other tasks, such as sowing, hoeing, weeding, manuring, milking, various assistance tasks etc.).

Multiple configurations of the model have been tested. As results show, it turned out that there was no single exclusive function meeting initial requirements, and hence, multiple directions had to be explored (as described in the next sections).

9.4.2   Crop Production Model

The purpose of this model is to compare agricultural strategies likely to be employed by the oppidum's population in relation to the necessary land-use area and ratio of the population engaged in agricultural work, in order to find out (1) whether the hinterland of the site itself had capacity to sustain constantly growing population of the oppidum and (2) if the oppidum's society could support food non-producers (craft specialists or elite). Following model inputs were defined (see Fig. 9.4):

- two time series from the population growth model (giving the values of the caloric requirements and available workforce for each year of simulation, cf. Sect. 9.4.1),
- the map of arable land around the settlement (modelled in GIS, cf. Sect. 9.3.3),
- the type of the agricultural strategy (cf. Sect. 9.3.4).

Three different land cultivation strategies (intensive–extensive–mixed) were implemented. While intensive farming provides higher yields on smaller area than extensive farming, it also requires significantly higher labour input. The production of cereals should be at least equal to the total consumption (plus seed corn— from every harvest 200 kg of grain/ha must be secured for the next-year's sowing) and should be achieved with the available workforce. Yields above the actual consumption represent a surplus and are stored. Around 10–15 % is accounted for losses. Keeping part of the surplus grain in storage until the next harvest can substantially diminish the impact of attested harvest fluctuations. Currently, these are driven randomly with the addition of sudden "events" (such as hailstorms, frosts or flash-floods). Each year there is a probability of some such event which can reduce the total harvest. Due to the absence of sunken silos, the maximum storage period was set to three years. After that time, the crop storage from the first year must have been consumed (surplus grain could be for example fed to

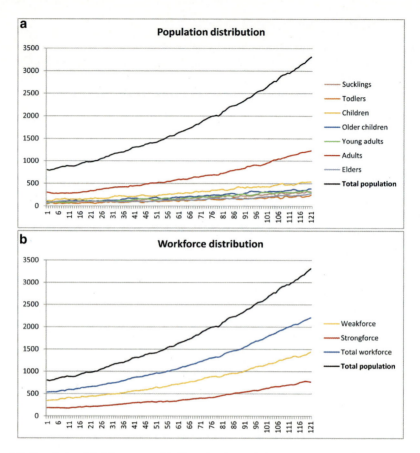

Fig. 9.3 Population Model outputs. (**a**) Growth of population—age groups, for initial number of 800 inhabitants. (**b**) Growth of strong, weak and total workforce, for initial number of 800 inhabitants

animals) or disposed of.[5] In case of a harvest failure, the oppidum's population should compensate using their reserves. This way, the years of bad harvests are counted as "bad-years" when the consummation level is higher than the production level and as "critical-years" when there are three bad years in a row. If the stored reserves are depleted as well, the model returns "years-with-no-food" which means that the population faces a crisis with acute food shortage.

The map of land suitability was imported to NetLogo, to be used as the background for visualization (Fig. 9.5). The possibilities are either to use a generalised raster map (as .png file) or to import an ascii text file. We used second option,

---

[5]This issue can be further examined for example also in relation to the feasting events when the surplus grain is consumed (cf. Van der Veen and Jones 2006).

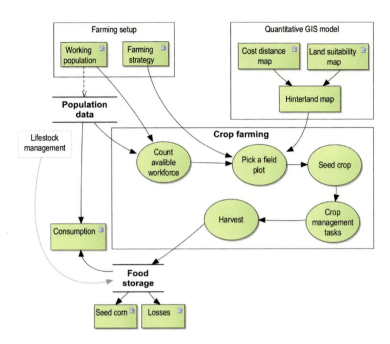

Fig. 9.4 Crop production model

because it preserved original coding of layers and realistic spatial proportions. The oppidum is situated in the middle of the map and the area around was classified according to the "utility index" of the suitability for cultivation (cf. Fig. 9.1).

The following model inputs are defined:

- the average crop and its standard deviation in case of intensive and extensive management (between 700–3,000 and 500–2,000 kg/ha respectively),
- the ratio ("strongforce", "weakforce) of working population (between 0 and 1),
- the ratio of cereal consumption (between 0 and 1),
- the number of workers per hectare for each strategy,
- seed corn for the following year and losses.

Listing 9.2 Crop production model pseudocode

```
model-setup
  load map
  update visualization
    recolor patches according to attractiveness, distance and
      ↪ current state of  cultivation
  load population data
  load consumption data
  load workforce data
  initiate animal-agents
end
```

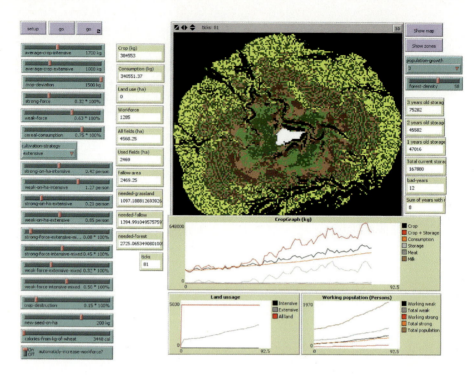

Fig. 9.5 Crop Production Model interface: sliders for setting initial values of parameters (*left*), visualization of the land-use in case of extensive strategy (*right*)

```
model-go
  repeat for 120 steps
      set global variables
         count number of animals
         count cereals requirement
      update fields allocation according strategy
      update crop
         count crop
         count storage
         count destroyed storage
      update animal data
         slaughter-animals
         milking
      update plots and monitors
      update visualization
         recolor patches according attractiveness, distance and
         ↪ current state of cultivation
end
```

The initialization of the model (refer to Listing 9.2) consists of loading and visualization of input data, especially loading text files with the time series of population, workforce and consumption. In the main procedure, the cycle repeats

120 times to simulate 120 years. In each simulated year, global variables are updated, characteristics of each patch of the land are updated according to current agricultural strategy and related crop data are processed and visualized.

We compared three agricultural strategies experimentally, with the aim to identify the appropriate labour input and carrying capacity of the hinterland. Frequencies of bad and critical years as well as years without food were examined, in order to ascertain under which model setups stress situations occur. The frequent appearance (i.e. several consecutive years) of bad, critical or even years-with-no-food means the particular agricultural strategy was not applicable.

With the outputs from the Population Model (cf. Sect. 9.4.1) the labour input under all strategies required between 25–40 % of the male and young male adults ("strong-force") and 40–55 % of the rest of the population ("weak-force") (see Fig. 9.6). That means that not all of the oppidum's population had to be engaged in the agricultural (meaning cereal production) work.

Other set of experiments was focused on the sustainability of the land-use in relation to strategies employed. In the case of an intensive strategy being employed, a population experienced several bad-years between the years 18 and 38 plus one period of critical-years (years 36–38) caused most likely by the labour shortage (returned from the Population Model). After that the production was stable to the ninth decade (year 93) of the oppidum's occupation, where the farming had to be carried out in more distant field plots with decreased net returns (Figs. 9.6a and 9.7a) because of the applied cost penalty. Then, the intensive strategy could not be efficiently practiced, because more labour input had to be invested into the necessary subsistence tasks. In total, a population of ca. 2,500 persons could sustain the agricultural production until the end of the oppidum's existence, but struggled considerably from the ninth decade.

In case of an extensive strategy being used, the simulated population did not experience problems with the labour shortage in the third decade, but from the year 92 onward, the population of ca. 2,000 people would encounter problems with availability of the arable land (see Figs. 9.6b and 9.7b). Due to extensive fallowing, the cultivated area would be much larger than with the intensive strategy. The cost penalty (though with moderate impact compare to the intensive strategy) also influenced the net returns from the fields. The population would increase for another 10–15 years, living on storage reserves, but after that period, the land-use approach must have changed, the crops must have been supplied externally, or the population density must have declined. The process of adaptation to the new conditions may have included the intensification of the agricultural practices, in order to reduce the cultivated area.

A combination of intensive and extensive cultivation practices appeared as a suitable compromise between land and labour shortages under constantly growing population density, though during the third decade the population experienced similar situation as when practising the intensive strategy (Fig. 9.6c). The experimental results showed that the population can maintain a more or less constant level of surplus (turned into storage), and must therefore have experienced a minimum of crisis situations due to harvest failures. In our experiments, the ratio of the

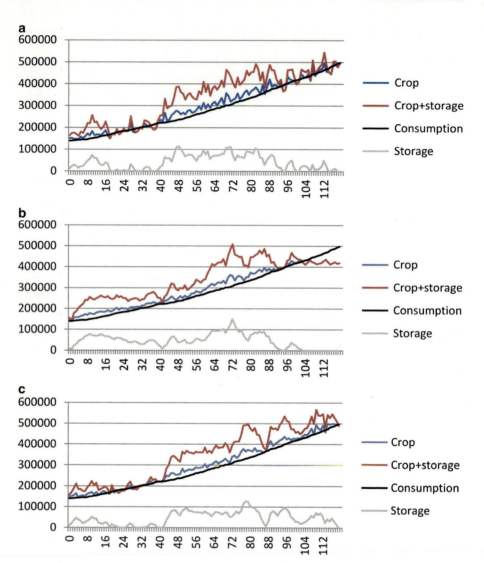

Fig. 9.6 Sustainability of production under different agricultural practices. (**a**) Intensive strategy: strongforce 34 %, weakforce = 55 %, average crop 1,500 kg/ha, standard deviation 1,000. (**b**) Extensive strategy: strongforce 25 %, weakforce = 40 %, average crop 1,000 kg/ha, standard deviation 500. (**c**) Mixed strategy: strongforce 29 %, weakforce = 50 %, average crop 1,000–1,500 kg/ha, standard deviation 700

intensive and extensive cultivation practices within the mixed method was set to 1/3:2/3 (intensive:extensive). However, choosing between the two strategies or their combination depends on the decisions of the farmers who are affected by many other factors (quality and accessibility of land, availability of the workforce, preference of

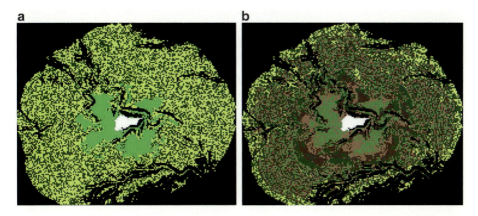

Fig. 9.7 Total field areas. (**a**) Intensive strategy: average crop 1,500 kg/ha with standard deviation 1,500, strong force 0.42 person/ha, weak force 1.27 person/ha. (**b**) Extensive strategy: average crop 900 kg/ha with standard deviation 1,000, strong force 0.21 person/ha, weak force 0.85 person/ha

other subsistence strategies like stock farming, climate changes, bad harvests etc.). Such a process is presented in the Field Allocation Model, using fuzzy methods as technical basis.

9.4.3 Field Allocation Model

While the typical process of human-driven research concerning the land-use is usually performed using GIS tools with multi-criteria decision analysis (Eastman 2006, pp. 126–134), our model provides an alternative approach with possible extensions in terms of household/people interactions with the landscape (cf. Olševičová et al. 2012; Machálek et al. 2013). In order to model the farmers' decision processes concerning spatial structure of their fields, we created fuzzy system helping to provide initial hypotheses in terms of crop field layouts.

Fuzzy rule-based systems build on top of the theory of fuzzy sets and fuzzy logic introduced by L. Zadeh in 1965. With classical sets, we assign each object either an "is-member" or an "is-not-member" property. On the contrary, in case of fuzzy sets, a *membership degree function* in the interval [0; 1] is used (see e.g. Babuška 2001; Ross 2010). This leads to a different (but still consistent with our understanding of Boolean logic) concept of set operations and related logical operations. Fuzzy rule-based systems provide a way to encode domain specific knowledge and control behaviour of a system or entity in conformity with this knowledge. The rules have following general form: IF *antecedent proposition* THEN *consequent proposition*.

Fuzzy propositions are statements like: "x is big", where "big" is a linguistic label, defined by a fuzzy set on the universe of discourse (Babuška 2001). Linguistic labels are also referred to as fuzzy constants, fuzzy terms or fuzzy notions.

A linguistic variable is a quintuple (Klir and Yuan 1995)

$$L = (x, A, X, g, m) \tag{9.5}$$

where x is the base variable (it also represents the name of the linguistic variable), $A = A_1, A_2, \ldots, A_n$ is the set of linguistic terms, X is the domain (universe of discourse) of x, g is a syntactic rule for generating linguistic terms and m is a semantic rule that assigns to each linguistic term its meaning (a fuzzy set in X).

Attempting to encode the knowledge of Iron Age farmers in terms of land suitability and accepting their limited analytical capabilities we see a fuzzy rule-based system containing a set of if-then rules as a natural tool to express their decision processes (i.e. subjective, approximate, using terms like "near", "far", "weak", "strong", "fair"; cf. Ross 2010).

This model simulates the farmers' decision-making process regarding suitability of individual land patches for crop (or animal) husbandry and also about correction of predefined percentages of intensive farming according to the difference between actual harvest and annual nutritional requirements of the community. In addition, the total harvest can be modified by a long term trend function to simulate progressively decreasing or increasing carrying capacity of the area.[6] Taking into account existing land suitability evaluation (cf. Sect. 9.3.1) we have split the problem into two levels:

- The first level represents an evaluation of land based only on the terrain's invariant properties (such as distance, slope, etc.). The problem how individual land characteristics influence suitability has already been addressed in human-landscape occupational rules as well as in the field of agricultural practices (cf. Jarosław and Hildebrandt-Radke 2009; Reshmidevi et al. 2009).
- The second level introduces a dynamic factor (e.g. the harvest from the previous season). An evaluation process from the first level can be understood here as an initial step (time "zero") where inhabitants have not yet influenced the environment in any way. But in the next season, their knowledge becomes broader, because they can evaluate the results of their previous assumptions on what part of the land is more or less suitable to be farmed with a specific cultivation strategy.

A single type of agent—a household which represents one or more families living in a settlement (house or a group of houses) with the arable land around, is defined. We assume that the "hinterland" is based on exploring accessible areas around the settlement, which lies in the centre.

Our proposed fuzzy inference requires four linguistic variables as input:

1. the distance of individual land patches from the household,

---

[6]This model forms a part of the main group of Agricultural Models. Its trial runs are presented on the smaller site (four households)—the lowland open settlement of Ptení, where the landscape settings are similar to the oppidum of Staré Hradisko.

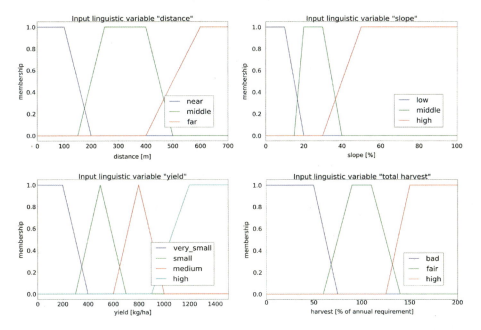

Fig. 9.8 Input variables of the fuzzy inference system

2. a slope gradient,
3. suitability of individual land patches,
4. the total harvest (as a percentage of inhabitants' annual nutrition requirements).

The concrete form of the membership functions related to these variables is the result of the previous GIS analysis (for the parameters see Sects. 9.3.1 and 9.3.3) and also of empirical testing (Fig. 9.8). A key parameter influencing both total and patch-level yields is the suitability of soil. The model works with 5 "soil categories": unfarmed soil of the settlement, alluvial soil and three additional qualitative categories for arable land.

The stochastic nature of crop yields required selecting a proper random distribution. While there exist objections against normally distributed crop yields (e.g. Ramirez et al. 2001), modelling yields by normal distribution still cannot be refused in general (e.g. Upadhyay and Smith 2005). We have applied it also in our model, due to the lack of detailed evidence. Coming to details, the mean of the distribution separates aforesaid yield ranges into equal halves and standard deviation is defined so that the maximum and minimum values are at $\pm 3\sigma$. The estimated distributions $N(\mu; \sigma^2)$ were calculated as:

$$\mu = \frac{(Y_{min} + Y_{max})}{2} \tag{9.6}$$

$$\sigma = \frac{(Y_{max} - Y_{min})}{6} \tag{9.7}$$

As an output of the fuzzy inference system, two linguistic variables have been defined:

1. suitability—this variable quantifies suitability of a single land patch in terms of its usability for growing crops. Patches with suitability near a value of 100 can be understood as very suitable (near the household, flat and with good yield potential). Patches near 0 are considered to be inappropriate (far lying, sloping, low yield potential).
2. intensity—although the model operates with a parameter, which specifies the percentage of household's arable land, it also provides auto-correction of this value according to the difference between required and actual harvest. The real proportion of cultivated land is calculated as a product of farming and intensity. The variable intensity is expected to be approximately 1 if the total harvest is about equal to the requested value (parameter required-annual-yield).

Listing 9.3 Fuzzy-rules pseudocode for The Field Allocation model

```
RULE 1:  IF yield IS very_small THEN suitability IS low;
RULE 2:  IF slope IS high THEN suitability IS low;
RULE 3:  IF distance IS near AND (slope IS low OR slope IS middle)
    ↪    AND (yield IS high OR yield IS medium) THEN suitability IS
    ↪ high;
RULE 4:  IF distance IS near AND (slope IS high) AND (yield IS
    ↪ high) THEN suitability IS middle;
RULE 5:  IF distance IS middle AND slope IS low AND (yield IS NOT
    ↪ very_small) THEN suitability IS high;
RULE 6:  IF distance IS middle AND slope IS middle AND yield IS
    ↪ high THEN suitability IS high;
RULE 7:  IF distance IS far AND (slope IS low) AND (yield IS
    ↪ medium OR yield IS high) THEN suitability IS middle;
RULE 8:  IF distance IS far AND (slope IS middle) AND (yield IS
    ↪ NOT very_small) THEN suitability IS low;
RULE 9:  IF distance IS near AND slope IS low AND yield IS NOT
    ↪ very_small THEN suitability IS high;
RULE 10: IF distance IS middle AND slope IS low AND (yield IS
    ↪ small OR yield IS very_small) THEN suitability IS low;
RULE 11: IF harvest IS fair THEN intensity IS normal;
RULE 12: IF harvest IS bad THEN intensity IS high;
RULE 13: IF harvest IS high THEN intensity IS low;
```

To calculate yield y_p of a specific patch we have defined the following function:

$$y_p = \frac{25 \cdot h \cdot y}{10000} \cdot T \cdot r$$

where $\frac{25}{10000}$ recalculates per-unit yield to model's patch size, h is a yield per hectare, y is normal random variable with properties so that $h \cdot y$ has properties of the distribution defined in Eqs. (9.6) and (9.7). T is a coefficient expressing an influence of a farming type (it has value of 1 for intensive farming and 0.3 for extensive farming) and r is an optional coefficient to apply long-term trends (i.e. climate change). For the fuzzy inference system, 13 rules have been defined; these can be found in Listing 9.3.

As always in fuzzy logic, the operator AND is defined using the min function, while OR is defined using max function. For rule activation, the min function was used. For rule accumulation, a bounded sum method (i.e. $min(1, \mu_A + \mu_B)$) was used. Fuzzy implication in based on Mamdani's inferencing scheme.

Figure 9.9 presents an example of land evaluation. Land area in the model is represented by a grid of discrete patches of the same size. Stochastic properties of the model cause that even several neighbouring patches may sometimes differ significantly in terms of calculated suitability. Such result is hard to interpret directly, because we cannot expect that a real farmer was mixing crop husbandry practices (including the "no-use" one) every few meters. To resolve this problem, we have at the moment proposed to post-process the resulting "suitability map" using linear filtering:

$$g(i, j) = \sum_{k,l} f(i + k, j + l)h(k, l) \tag{9.8}$$

Here, f represents the input signal, h is called linear convolution kernel. We have been using the following bilinear kernel:

$$h(k, l) = \frac{1}{16} \begin{pmatrix} 1 & 2 & 1 \\ 2 & 4 & 2 \\ 1 & 2 & 1 \end{pmatrix} \tag{9.9}$$

The suitability map can be understood as a spatial signal in which a proper linear convolution kernel serves as a low-pass filter which attenuates higher frequencies from the signal (Szeliski 2010).

While treating the suitability as a continuous variable is convenient in terms of described calculations, it is more complicated to apply such a variable directly in practice (i.e. what should farmer do exactly if the suitability is x?). Because of that we defined a mapping from continuous suitability variable to an ordinal set of suggested farming types (see Fig. 9.9).

Figure 9.10 shows a comparison between discussed fuzzy model and multi-criteria based solution which considers only distance and soil type. Given a required target yield, we can see that the fuzzy model is capable of producing results with comparable accuracy, the only exception being low intensity farming: the model evolves from a defined starting property size and is unable to reach the required area within the running time limit (it evaluates too many patches as unsuitable).

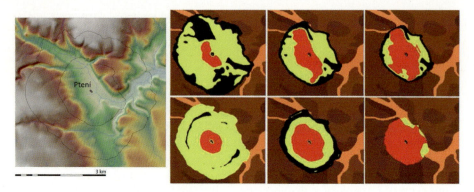

Fig. 9.9 The open settlement of Ptení: fuzzy (*first row*) and multi-criteria based (*second row*) model, required annual yield = 9,000 kg, required ratios for intensive farming = 10 % (*left*), 50 % (*centre*) and 90 % (*right*), *red area* = best suitability for intensive farming, *yellow area* = best suitability for extensive farming or pastures, *black area* = evaluated land of lower suitability

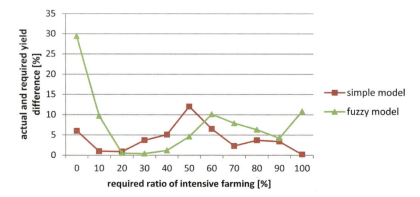

Fig. 9.10 Fuzzy (including linear filtering) and alternative model—target yield accuracy

9.5   Discussion

Results achieved can be discussed in the light of the framework of available data: according to the archaeological record, the settlement density in the late La Tène period having increased over some time and then decreased again rapidly. The goal of our modelling effort was to ascertain whether "crossing the carrying capacity" situation was a main factor in the process of the abandonment of settlements. Setting the proper values was a difficult undertaking. We have only limited data concerning the initial and final populations at Staré Hradisko. Rather, we have information on the occupational development. Despite that, the Population dynamics model provides realistic time series of consumption, workforce and age distributions of population of the oppidum's agglomeration. It is essential to note, however, that its development did not take into account a sudden decline in the occupation observed in the archaeological record. This is because our goal was to explore the cause of this decline. To be able to address that issue, we first needed an invariable model of the population growth. Therefore, the modelling results showed a constantly growing population from the beginning until the end of the occupation (i.e. the "Baseline"). Maximum supposed "real" population density of the oppidum, reached only by natality, could then be seen around the years 80–90 of the simulation. With such a demographic profile, the oppidum's community could in fact practice all land-use strategies without any substantial problems apart from those imposed by natural harvest fluctuations due to weather and other (e.g. socio-economic) factors.

Using our models, we have proven by experiments that not all of the oppidum's population had to be engaged in the agricultural work. When other labour tasks are implemented (such as animal production and forest management), further experiments will be able to answer the question of sustainability of the non-producers at the oppidum. Since there is an archaeological evidence of elite members, which, presumably, were not involved in the agricultural production, the labour shortage may point to the necessity of using the external supplies.

The limits of the land-use strategies returned from the Crop Production Model, when the population was expected to react by adjusting their economic strategies, started acting around the population density being over 2,500 (under the intensive strategy—cost distance factors) and around 2,000 (under the extensive strategy—depleting of hinterland area). If the population growth would have reached this maximum value after first 80–90 years (due to massive immigration for example), that could be the realistic interpretation of the occupation's decline.

According to historical sources, exceeding the appropriate carrying capacity was not a rare occasion in history (cf. Schreg 2011, p. 312). Intensification led to innovations in the agriculture on the one hand, but also to a more rapid depletion of the land resources where their extent was limited on the other. When experiencing population growth, the households had to work harder in order to keep their life standards, due to the law of diminishing returns (Turchin and Nefedov 2009, p. 1). This concept refers to the fact that while the population increases exponentially, the growth of subsistence resources is only linear and generally slower. As the population approaches the carrying capacity, the production level gradually declines. This means that when reaching the carrying capacity threshold, the surplus becomes zero, and upon further population growth it becomes negative. At this point, the population faces a lack of resources for its reproduction and its density must decline (Turchin and Nefedov 2009, pp. 8, 10, Fig. 1.1a) or their subsistence strategies must be adapted to the new situation. Such adaptability processes may include changing the extensive cultivation practices into more intensive land use regimes (i.e. cultivating land with higher labour input on smaller area) or a change in economic preferences to stock farming or craft industries. The results of the mixed land-use model showed that the optimal strategy would have been to combine different agricultural practices. This inspired the Agricultural model II, which focuses on the decision process of farmers in choosing the optimum farming strategy in order to balance potential problems, with space and labour availability.

The results show that, since the adequate oppidum population (living from their own resources) was able to exploit environmental resources around the oppidum without simultaneously exhausting them, a rapidly growing number of inhabitants could—at some point—cross the limits of the sustainable agricultural production and experience several stress situations. Especially for the labour intensive scenario, the model resulted in diminishing returns from the cultivated land as the cost of farming more distant field increased. Around the population peak, the gradual depletion of stored reserves resulted in a supply crisis, which must have provoked a strong social response. This evolution can be a typical development towards societal decline following the distinctive upsweep accompanied by rapid population growth especially in the environment where the market economy was weak (Chase-Dunn et al. 2007; Turchin and Nefedov 2009). This hypothesis is a main subject for testing in further research.

9.6   Conclusion

This chapter has attempted to discuss the applicability of agent-based social simulation as a tool for the exploration of the late Iron Age oppida agglomerations, especially from the point of view of the population growth and related sustainability of production. By presentation of three consecutive NetLogo models—the model of population dynamics and two agricultural models—we intended to demonstrate the ability to move from a static data set (archaeological and environmental records) to dynamic modelling that incorporates feedback mechanisms, system integration, and nonlinear responses to a wide range of input data.

Even in case when detailed data is limited, these models could point to the constraints of the particular agricultural strategy and population density in relation to the specific environment (Schreg 2011, p. 307). In our case, evolution of computational techniques such as genetic algorithms (available through BehaviorSearch) or fuzzy rules (implemented in our jFuzzy plug-in) helped us to identify missing values of parameters and to optimize model settings. This approach can help to analyse past socio-economic processes, determine possible crisis factors and understand ecological and cultural changes.

Future studies will build upon the presented models. The applied approach can be adapted for other regions, and other economic strategies can be explored. Our next objectives are to investigate further the population structure of the oppidum, by incorporating different types of households and related attributes of the individuals. The agricultural models are planned to be enhanced by more detailed weather data, analyses of the animal production and related labour input resulting into a more complex image of the social structure. The models of interaction within the region will focus on the possibility of the food and raw resources circulation through social contacts. Also, elaboration of the decision process of Iron Age farmers promises an encouraging way where to direct our further research. At the moment, our models end with the limits of the given agricultural practice, when reaching a certain population density. By including the social variables representing farmers' independent decisions to change from one economic strategy into another (or to adopt new ones) in order to cope with worsening conditions of the sustainable agricultural practice, our model can approach the past social complexity studies on a new level.

Acknowledgements This paper was made possible by the "Social modelling as a tool for understanding Celtic society and cultural changes at the end of the Iron Age" project, supported by the Czech Science Foundation (Grant No.P405/12/0926), and by the "Agent-based models and Social Simulation" project, supported by the Grant Agency of Excellence, University of Hradec Králové, Faculty of Informatics and Management. The authors would like to thank the editors and the anonymous reviewers for their helpful and constructive comments that greatly contributed to improving the final version of the chapter.

References

Babuška R (2001) Fuzzy and Neural Control Disc. Course Lecture Notes, Delft University of Technology
Baili P, Micheli A, Montanari A, Capocaccia R (2005) Comparison of four methods for estimating complete life tables from abridged life tables using mortality data supplied to EUROCARE-3. Math Popul Stud 12:183–198
Bakels C (2009) The Western European Loess Belt: Agrarian History, 5300 BC – AD 1000. Springer, Leiden
Bayliss-Smith T (1978) Maximum Populations and Standard Populations: The Carrying Capacity Question. In: Green D, Haselgrove C, Spriggs M (eds) Social Organisation and Settlement: Contributions from Anthropology, Archaeology, No. 47(i) in BAR International Series (Supplementary), Oxford, pp 129–151
Brázdil R, Valášek H, Chromá K (2006) Documentary evidence of an economic character as a source for the study of meteorological and hydrological extremes and their impacts on human activities. Geogr Ann A 88(2):79–86
Bryson RA, DeWall K (eds) (2007) A Paleoclimatology Workbook: High Resolution, Site-Specific, Macrophysical Climate Modeling. Mammoth Site of Hot Springs, Hot Springs
Campbell BMS (2007) Three Centuries of English Crops Yields, 1211–1491. Medieval Crop Yields Database, The School of Geography, Archaeology, and Palaeoecology, The Queen's University of Belfast. http://www.cropyields.ac.uk
Chase-Dunn C, Niemeyer R, Alvarez A, Inoue H, Sheikh-Mohamed H, Chazan E (2007) Cycles of Rise and Fall, Upsweeps and Collapses: Changes in the Scale of Settlements and Polities Since the Bronze Age. In: Conference on Power Transitions, University of Indiana, Bloomington. http://irows.ucr.edu/papers/irows34/irows34.htm
Chisholm M (1979) Rural Settlement and Land Use. An Essay in Location. Hutchinson, London
Cingolani P, Alcalá-Fdez J (2012) jFuzzyLogic: A Robust and Flexible Fuzzy-Logic Inference System Language Implementation. In: Proceedings of the IEEE World Congress on Computational Intelligence 2012
Čižmář M (2005) Keltské oppidum Staré Hradisko. No. 4 in Archeologické památky střední Moravy, Vlastivědné Muzeum v Olomouci, Olomouc
Danielisová A (2006) To the Architecture of Oppida: Reconstruction of One Part of Settlement Pattern Behind the Ramparts. In: Architektura i budownictwo epoki brązu i wczesnych okresów epoki żelaza: Problemy rekonstrukcji. Muzeum Archeologiczne, Biskupin, pp 269–302

Danielisová A, Hajnalová M (in print) Oppida and Agricultural Production - State of the Art and Prospects: Case Study from the Staré Hradisko Oppidum (Czech Republic). In: Produktion – Distribution – Ökonomie. Siedlungs- und Wirtschaftsmuster der Latènezeit. Kolloquium Otzenhausen 2011, Universitätsforschungen zur Prähistorischen Archäologie, S. Hornung

Del Monte-Luna P, Brook BW, Zetina-Rejón MJ, Cruz-Escalona VH (2004) The carrying capacity of ecosystems. Glob Ecol Biogeogr 13:485–495

Dreslerová D (1995) A Settlement - Economic Model for a Prehistoric Microregion. Settlement Activities in the Vinoř-Stream Basin During the Hallstatt Period. In: Kuna M, Venclová N (eds) Whiter Archaeology? Papers in Honor of Evžen Neústupný, Institute of Archaeology, Prague and Academy of Sciences of the Czech Republic, pp 145–160

Eastman JR (2006) IDRISI Andes: Guide to GIS and Image Processing. Clark University, Worcester. http://clarklabs.org/terms-of-use.cfm

Ebersbach R (2002) Von Bauern und Rindern: Eine Ökosystemanalyse zur Bedeutung der Rinderhaltung in bäuerlichen Gesellschaften als Grundlage zur Modellbildung im Neolithikum. No. 15 in Basler Beiträge zur Archäologie, Schwabe, Basel

Erdkamp P (2005) The Grain Market in the Roman Empire: A Social, Political and Economic Study. Cambridge University Press, Cambridge

Fischer E, Rösch M, Sillmann M, Ehrmann O, Liese-Kleiber H, Voigt R, Stobbe A, Kalis AJ, Stephan E, Schatz K, Posluschny A (2010) Landnutzung im Umkreis der Zentralorte Hohenasperg, Heuneburg und Ipf: Archäobotanische und archäologische Untersuchungen und Modellberechnungen zum Ertragspotential von Ackerbau und Viehhaltung. In: Krausse D (ed) "Fürstensitze" und Zentralorte der frühen Kelten, Band 2, no. 120 in Forschungen und Berichte zur Vor- und Frühgeschichte in Baden-Württemberg, Theiss, Stuttgart, pp 195–265

Gorenflo LJ, Gale N (1990) Mapping regional settlement in information space. J Anthropol Archaeol 9:240–274

Gregg SA (1988) Foragers and Farmers. Population Interaction and Agricultural Expansion in Prehistoric Europe. University of Chicago Press, Chicago

Hajnalová M (2003) Rastlinné makrozvyšky zo Starého Hradiska. Technical Report, Výskumná správa archeobotanická

Halstead P (1987) Traditional and ancient rural economy in Mediterranean Europe: plus ça change? J Hell Stud 107:77–87

Halstead P (1995) Plough and power: the economic and social significance of cultivation with the ox-drawn ard in the Mediterranean. Bull Summerian Agric 8:11–21

Halstead P, Jones G (1989) Agrarian ecology in the Greek islands: time stress, scale and risk. J Hell Stud 109:41–55

Halstead P, O'Shea J (eds) (1989) Bad year economics. Cambridge University Press, Cambridge

Hejcman M, Kunzová E (2010) Sustainability of winter wheat production on sandy-loamy Cambisol in the Czech Republic: results from a long-term fertilizer and crop rotation experiment. Field Crops Res 115:191–199

Higgs ES, Vita-Finzi C (1972) Prehistoric Economies: A Territorial Approach. In: Higgs E (ed) Papers in Economic Prehistory. Studies by Members and Associates of the British Academy Major Research Project in the Early History of Agriculture. Cambridge University Press, Cambridge

Jarosław J, Hildebrandt-Radke I (2009) Using multivariate statistics and fuzzy logic system to analyse settlement preferences in lowland areas of the temperate zone: an example from the polish lowlands. J Archaeol Sci 36:2096–2107

Klir G, Yuan B (1995) Fuzzy Sets and Fuzzy Logic: Theory and Applications. Prentice-Hall, Englewood Cliffs

Kohler TA, van der Leeuw SE (eds) (2007) The Model-Based Archaeology of Socionatural Systems. School for Advanced Research Resident Scholar Book, Santa Fe

Kunzová E, Hejcman M (2009) Yield development of winter wheat over 50 years of FYM, N, P and K fertilizer application on black earth soil in the Czech Republic. Field Crops Res 111:226–234

Küster H (1993) The Carbonized Plant Remains. In: Wells P (ed) Settlement, Economy and Cultural Change at the End of the European Iron Age. Excavations at Kelheim in Bavaria, 1987–1991, International Monographs in Prehistory, Archaeological Series, vol 6, Ann Arbor, pp 57–60

Lindsay J (2012) Whitebox Geospatial Analysis Tools. Manual. http://www.uoguelph.ca/~hydrogeo/Whitebox/Help/MainHelp.html

Machálek T, Cimler R, Olševičová K, Danielisová A (2013) Fuzzy Methods in Land Use Modeling for Archaeology. In: Proceedings of Mathematical Methods in Economics 2013

Meduna J (1961) Staré Hradisko. Katalog nálezůuložených v muzeu města Boskovic. FAM II. Brno. Museum Catalogue

Meduna J (1970a) Das keltische oppidum Staré Hradisko in Mähren. Germania 48:34–59

Meduna J (1970b) Star'e Hradisko, ii. Katalog der Funde aus den Museen in Brno (Brünn), Praha (Prag), Olomouc, Plumlov und Prostějov. FAM V. Brno. Museum Catalogue

Neuhäuslová Z (2001) Potential Natural Vegetation of the Czech Republic, Braun-Blanquetia: recueil de travaux de géobotanique, vol 30. Dipartimento di Botanica ed Ecologia dell'Università di Camerino, Camerino, map

Olševičová K, Cimler R, Machálek T (2012) Agent-Based Model of Celtic Population Growth: NetLogo and Python. In: Nguyen N, Trawiński B, Katarzyniak R, Jo GS (eds) Advanced Methods for Computational Collective Intelligence. Studies in Computational Intelligence, vol 457. Springer, Berlin

Ramirez O, Mishra O, Field J (2001) Are Crop Yields Normally Distributed? Paper Presented at the Annual Meeting of the American Agricultural Economics Association, Chicago

Reshmidevi T, Eldho T, Jana R (2009) A GIS-integrated fuzzy rule-based inference system for land suitability evaluation in agricultural watersheds. Agric Syst 101(1–2):101–109

Ross T (2010) Fuzzy Logic with Engineering Applications, 3rd edn. Wiley, Chichester

Rothamsted Research (2006) Guide to the Classical and Other Long-Term Experiments, Datasets and Sample Archive. Lawes Agricultural Trust, Bury St. Edmund

Russell K (1988) After Eden. The Behavioral Ecology of Early Food Production in the Near East and North Africa, BAR International Series, vol 391. British Archaeological Reports, Oxford

Salač V (2000) The Oppida in Bohemia: A Wrong Step in the Urbanization of the Country? In: Guichard V, Urban SSO (eds) Les Processus d'urbanisation à l'âge du Fer, Collection Bibracte 4, Archaeological Centre Bibracte, Glux en Glenne, pp 151–156

Salač V (2006) Die keltischen Oppida und ihre Macht. In: Krenn-Leeb A (ed) Wirtschaft, Macht und Strategie – Höhensiedlungen und ihre Funktionen in der Ur- und Frühgeschichte, Archäologie Österreichs Spezial, vol 1, Österreichische Gesellschaft für Ur- und Frühgeschichte, pp 233–245

Saller R (1994) Patriarchy, Property and Death in the Roman Family. Cambridge Studies in Population, Economy and Society in Past Time. Cambridge University Press, Cambridge

Schreg R (2011) Feeding the Village - Reflections on the Ecology and Resilience of the Medieval Rural Economy. In: Klápště J, Sommer P (eds) Processing, Storage, Distribution of Food. Food in the Medieval Rural Environment (Ruralia VIII), Ruralia, vol VIII. Brepols, Turnhout, pp 301–320

Stonedahl F (2011) Genetic Algorithms for the Exploration of Parameter Spaces in Agent Based Models. Ph.D. thesis, Northwestern University, Evanston. http://forrest.stonedahl.com/thesis/forrest_stonedahl_thesis.pdf

Stonedahl F, Wilensky U (2010) Evolutionary Robustness Checking in the Artificial Anasazi Model. In: Proceedings of the 2010 AAAI Fall Symposium on Complex Adaptive Systems

Szeliski R (2010) Computer Vision: Algorithms and Applications. Springer, Berlin. http://szeliski.org/Book/

Turchin P (2009) Long-term population cycles in human societies. Ann N Y Acad Sci 1162:1–17

Turchin P, Nefedov SA (2009) Secular Cycles. Princeton University Press, Princeton

Upadhyay B, Smith E (2005) Modeling Crop Yield Distributions from Small Samples. In: Canadian Agricultural Economics Society Annual Meeting, San Francisco. http://purl.umn.edu/34161

Van der Veen M, Jones G (2006) A re-analysis of agricultural production and consumption: implications for understanding the British Iron Age. Veg Hist Archaeobot 15:217–228

Wainwright J (2008) Can modelling enable us to understand the rôle of humans in landscape evolution? Geoforum 39:659–674

Wilensky U (1999) NetLogo. Center for Connected Learning and Computer-Based Modeling, Northwestern University, Evanston. http://ccl.northwestern.edu/netlogo/

Wright HT (2007) Progress in Cultural Modeling. In: Kohler T, van der Leeuw SE (eds) The Model-Based Archaeology of Socionatural Systems. SAR Press, Santa Fe, pp 229–232

Chapter 10
Simulating Patagonian Territoriality in Prehistory: Space, Frontiers and Networks Among Hunter-Gatherers

Joan A. Barceló, Florencia Del Castillo, Ricardo Del Olmo, Laura Mameli, Francisco J. Miguel Quesada, David Poza, and Xavier Vilà

10.1 Introduction

"Ethnicity", "territoriality" and "culture" are still fashionable words in modern archaeological research. Maybe such popularity is a signal of an academic inertia that has kept some of the narrative of the old historical and cultural traditions, having varied the background of the narrative. Although the current effort to develop an archaeology of identity and ethnicity is impressive, there are still many questions to be solved and even asked regarding to the role of archaeology and archaeological data in dealing with such concepts.

The proper question is "*why* groups of people were the way they were in the past"—That means, *how* social aggregates emerged. The complex interplay of people and their social actions, and the consequences of those actions would explain ethnicity in terms of a vast network of interacting *actions* and *entities*. We assume that ethnogenesis and identity formation emerged among prehistoric hunter-gatherers as result of the contradiction between social inertia (knowledge

J.A. Barceló (✉) • F.D. Castillo • F.J.M. Quesada • X. Vilà
Universitat Autónoma de Barcelona, Barcelona, Spain
e-mail: juanantonio.barcelo@uab.cat; florenciadelcastillo@hotmail.com;
miguel.quesada@uab.cat; xavier.vila@uab.cat

R.D. Olmo
Universidad de Burgos, Burgos, Spain
e-mail: rdelolmo@ubu.es

L. Mameli
Universitat de Girona, Girona, Spain
e-mail: lauramameliiriarte@gmail.com

D. Poza
Universidad de Valladolid, Valladolid, Spain
e-mail: djpoza@gmail.com

© Springer International Publishing Switzerland 2015   217
G. Wurzer et al. (eds.), *Agent-based Modeling and Simulation in Archaeology*,
Advances in Geographic Information Science, DOI 10.1007/978-3-319-00008-4__10

inheritance) and cultural consensus (social similarity) built during cooperation and labor exchange.

In this paper we consider Patagonian historical trajectory as a case study. At the extreme south of South America, hunter-gatherers survived until European expansionism altered tragically 13,000 years of history. In the last 40 years, the very idea of a "Patagonian" ethnicity has evolved from a static and essentialist classification of human groups according to their very own *nature*, to a relational frame of reference used by a group of people considered to be *similar* and differentiated from others (Del Castillo et al. 2011). We have adopted here an analytical view of ethnicity based on the emergence of identity as a consequence of the very fact that some individuals interact more often than others, which means that people embedded in social networks interact with a subset of the population and define themselves in terms of their similarity (or *cultural likeness*) to the people with whom they interact.

In this paper we have computationally grown a surrogate of hypothetical ethno-genetic processes that may have occurred in prehistoric Patagonia, to explore their effects on the emergence of "cultural" differences, spatial mobility and diffusion of innovations. Our aim has been to simulate virtual social agents "living" in a virtual environment defined on the basis of social theory and/or historical data. In this preliminary and simplified computer simulation we are exploring the consequences that labor exchange and territorial mobility in an artificial unconstrained world had on identity formation and negotiation. We expect to be able to discern if cultural diversity emerged as a result of social decisions only, or if it was the result of constraints on mobility generated by geography and the irregular distribution of resources, both in space and time. By implementing social events as computational agents and their mutual influences as interactions, we seek to discover whether collective action may be described and explained as non-accidental and non-chaotic. It should be emphasized that the aim of such work is *not* to create the most "realistic" artificial society possible. The simulation is not intended to be an exact re-creation of the past but rather provide us with an understanding of how different circumstances might have affected people in the past.

An important aspect of this way of understanding historical causality is that it forces the analysis to pay attention to the flux of ongoing activities, to focus on the unfolding of real activity in a real historical setting.

10.2   From Ethnicity to Territoriality

What has traditionally been called "ethnic" differentiation is nothing more than a consequence of the diverse degrees of social interaction between human communities, and an emerging pattern of social "similarity". We want to explore social mechanisms and processes whose results may produce some similarity in social activity. Developing the definitions of D'Andrade (1987), Carley (1991), Axelrod (1997), Boyd and Richerson (2006), and others, we may define culture as the distribution of information (ideas, beliefs, concepts, symbols, technical knowledge, etc.) capable of affecting individuals as a result of their interactions in the present. Cultural consensus is a measure of similarity in motives, actions, behavior and mediating artifacts that does not exist for ever, but is in the process of continuous building, influenced by many aspects of social life. Then, observed aggregations of activities and social practices may be good estimates of "culture", but the accuracy of that observation depends on the agreement among what people did, what they believed they did, and the number of observations on past actions (Romney 1999; Romney and Weller 1984; Romney et al. 1986; Garro 2000; Weller 2007; Sieck 2010; Borgatti and Halgin 2011).

If "culture" can be defined as the expected variance in a distribution of social values, goals and activities among synchronous human aggregates or populations, "ethnicity" can be approached as the degree of social inertia or resilience between different temporal states of the same aggregate or population, that is the ability of an aggregate of social agents to maintain a certain identity in the face of historical change and external perturbation (Stein 1997; Ramasco 2007; Castellano et al. 2009). Consequently, "culture" and "ethnicity" can also be understood as the propensity or tendency a human group has to practice or produce a distinct social goal, motivation, behavior or artifact. Both "culture" and "ethnicity" are quantitative properties of human aggregates and not features of individuals. Whereas "culture" expresses the degree of commonality in social activities between contemporaneous groups, "ethnicity" expresses the degree of similarity between social activities between different temporal stages of the same group. Therefore, we don't consider ethnic groups to be discontinuous isolates to which people naturally or "ideally" belong but a series of real related dichotomizations of inclusiveness and exclusiveness resulting from social reproduction, affecting the way people aggregated in the past and aggregate in the present into groups and adapted/adapt their social practices as a consequence. Ethnicity dos not presuppose the existence of discrete and particular "ethnia", nor does culture imply the existence of cultures. In other words, there is not a thing or a set of things called an "ethnic group", in the same way that there is not a set of things called "culture". The use of both terms should then be limited to the configuration of an instrument for measuring typified ideas, behaviors, actions and products that different human aggregates may have in common, in the present or across time.

Our starting point for the computer simulation of ethnogenetic processes is that the lesser the intensity and frequency of inter-group relationships, the greater the differences in ways of speaking and other *cultural features manifested by groups* (Del Castillo 2012; Del Castillo and Barceló in press). Commonalities in needs, motivations, goals, actions, operations, signs, tools, norms, cooperative ties, and in division of labor schemas are the consequences of the way some social agents interacted, aggregated in space and time as a consequence of some of these interactions, and reproduced the basis of such an aggregation. The formation of such diverse aggregates at diverse scales and with different degrees of similarly acting social agents is mediated by a perceived similarity both at the moment of the interaction, but also previously, as some inherited social inertia or resilience. As a result: the greater the temporal depth of the social aggregation of agents, and the longer the transmission links between a greater number of generations, the more redundant the consensus, and the more stable it is.

The obvious result is that most people with the same history of interactions show a degree of similarity in their motivations, goals, actions behaviors and mediating artifacts which do not depend on their actual will, but on what they have received from the past (Dow et al. 1984; Eff 2004). Endogamy appears then as one of the main factors to classify some groups of similar subjects, needs, motivations, behaviors and/or artifacts as ethnic (Abruzzi 1982; Whitmeyer 1997). Human reproduction is not just a mere biological process, but a socially mediated mechanism. Reproductive mates are consciously chosen and many social, ideological and political constraints impose some directionality in social reproduction (Bernardi 2003; Bongaarts and Watkins 1996; Kalick and Hamilton 1986). For that reason we assume social reproduction, i.e., the historically variable forms of mating and marriage, and kinship topologies have a key importance for the definition and analysis of ethnically distinguishable populations. Similarity and difference, continuity and distinction among local populations emerge when unions among members of a same population occur more often than unions with members of other populations. Endogamy preserves the differential distribution of similar subjects, needs, behaviors and artifacts inside a community. Exogamy, on the other hand, may attenuate local distinctions when syncretic knowledge and norms are transmitted from a generation to the following one. The more institutionalized the mechanisms of reproductive isolation—as part of the explicit norms of a community, constituting the law, religious doctrine or ideology of government-, the greater the possibility that similarity emerges and is consciously sought after and maintained in the community, and transmitted to the following generation (Abruzzi 1982; Cavalli-Sforza et al. 1994; Giuliano et al. 2006).

When useful commonality and redundancy emerges into a social encoding scheme, members of the group produce the means whereby descendents will *predict* their own identity even before acting, augmenting the social group and cognitive resilience and hence generalizing what defines their own group. The obvious consequence is a higher probability of social aggregation, which in turn increases the probability of within group regularity. We suggest the probability of an interaction between two agents is based on the principle of homophily, i.e. the greater the amount of knowledge they share the more probable the interaction is. And the more similar the origins, the more commonalities the members of the same descent group share. But what comes from the past can be modified in the present. Within the group and between groups, social agents may interact for many reasons and in many ways: cooperating to acquire subsistence, cooperating to produce (Bjerck and Zangrando 2013; Borrero et al. 2009; Gómez Otero 2007; Orquera and Gómez Otero 2007) tools and instruments, cooperating to exchange subsistence and/or tools, cooperating for reproducing themselves, refusing such cooperation, or compelling other agents to work in their own benefit, etc. War and conflict are also another kind of interaction. In all those cases, interactions vary in intensity and frequency, defining a complex network of positive or negative intergroup relationships. As a result, agents adopt similar activities, and their actions tend to generate similar results. Consequently, inter-action should be considered both positive (cooperation) and negative (conflict), in such a way that different kinds of social fusion, fission and friction develop a set of representations and values that set the terms from which social clustering and self-ascriptions are constructed. All these diverse sources of interaction modify similarity in social activities and any kind of perceivable regularity, in such a way that newly configured aggregates affect future mechanisms of social reproduction, and hence vary elements to be transmitted to the next generation. Social inertia changes constantly because social conditions in which reproduction takes place changes according to actual interactions. As a consequence, ethnicity, the degree of similarity product of social inertia may change constantly.

Our hypothesis is based on the idea in terms of agents' tendency to interact with agents with a similar "identity" which makes for a greater probability interactions between already connected people than unconnected ones (with dissimilar features). In addition, we also introduce the principle of social influence (i.e., the more people interact with similar people, the more similar they become) which runs at the level of communication and the formation of a socio-cognitive level.

10.3 Beginning of Times at the End of the World: Patagonia

How might those processes have acted in Patagonia before European colonization?

The antiquity of human settlement on the Patagonian steppe during the Pleistocene-Holocene transition is well established (Borrero 2008, 2012; Miotti and Salemme 2003; Mancini et al. 2013; Rivals et al. 2013). The beginning of human settlement in Patagonia has been reconstructed as a slow process of exploration and colonization, carried out by small groups, very mobile and dispersed, with approximate site-catchment areas of around 100 km in radius (Borrero and Barberena 2006; Barberena 2008). Sites may have been occupied intermittently, by few people and/or short periods of time, but repeatedly (Frank 2012). These foundational groups were characterized by low population density and the absence of specialized use of the ecosystem given the lack of competition among distant and dispersed groups (Barceló et al. 2009; Del Castillo 2012; Del Castillo and Barceló in press).

The growing number of sites identified as settlements and the greater rate of material deposition at those sites suggest a demographic increase during the Holocene (Borrero and Franco 1997; Borrero 2005; Martínez et al. 2013). It is from this time, around the seventh millennium B.P., that an increasing use of marine and littoral resources has been recorded (Bjerck and Zangrando 2013; Borrero et al. 2009; Gómez Otero 2007; Orquera and Gómez Otero 2007; L'Heureux and Franco 2002; Moreno et al. 2011). Many factors should be taken into account in understanding why different groups decided to manage differently diverse sets of resources from differentiated areas (Prates 2009). Among those factors, we can mention: the quantity of available labour, the size of the group, the degree of technological investment, the complexity of labor operations, the organizational needs of labor, and the social relations of production.

Six thousand years ago economic variability would have been consolidated all over Patagonia, defining a differentiation between some communities specialized in the exploitation of marine resources, some specialized in terrestrial resources, and others without specialization but exploiting both terrestrial and littoral resources. In any case, such economic differentiation cannot be well understood without taking into consideration mobility across the territory. Human groups moved from place to place for social and political needs, in such a way that extremely long and complex interaction networks developed. Goods and information would have traveled more than people would (Bellelli et al. 2008; Barberena et al. 2011; Méndez et al. 2012; Silveira et al. 2010; López et al. 2010). Borrero et al. (2008) explain the circulation of goods associated with large partially overlapping ranges within a framework of inter-group visits, which would be related to economic and social factors that make the size and composition very fluid over time.

If economic variability was already configured some time around 6,000/5,000 B.P., it is possible that cultural and social variability were also configured at the same time. How many "distinct" populations existed then? Ethnologists have made reference to language differentiation to suggest the existence of differentiated populations, even in these remote times. Languages spoken by hunters exploiting the forests along the Andes Mountains, the steppe, and the Atlantic littoral resources historically derive from two different linguistic sources originating around the same time Campbell (1997); Adelaar and Muysken (2004); Fernandez Garay (1998, 2004); Fabre (2005); Viega Barros (2005); Brown et al. (2008); Müller et al. (2009). The first one would have been spoken by human groups at the western side of the Andes (*mapudungun* linguistic family) (Zúñiga 2006) and another for human groups at the eastern side (*gününa a iajüch* and *chon-tsoneka*,[1] different languages from a common linguistic family) (Casamiquela 1983; Fernandez Garay 1998; Viega Barros 2005; Orden 2010). Further south and notably southwest, on the Pacific coast, canoe fishers spoke languages from a totally different origin: *chono-kawescar-yamana/yaghan* (Aguilera Faúndez 2007).

Nevertheless, cultural variability cannot be limited to the level of linguistic differentiation. It is important to remark that the speakers of those languages in early colonial times did not have ethnonyms for self identification: the term "gününa-küna", where "küna" is the word for "people" has been translated as "people of the same blood, friends" (Harrington 1946; Casamiquela 1983); "aonik'kenk", where "kenk" is the word for "people" has been translated as "people from the south" (Fernandez Garay 2004). There are many ethnographic sources on the inter-ethnic relationships in early colonial times (Nacuzzi 1998; Vezub 2006, 2009, 2011).

---

[1] Musters (1872–1873) wrote that "tsoneka" was the etnonym these people gave to themselves (Musters 1964 [1872–1873]). He was mistaken, because the work derives from *aónik'o ais* language: *tshontk' = ch'óon(e)k(e)* (Casamiquela 1965, p. 22).

There is also mention of mixed groups formed between coastal gatherers and interior hunters (*guaicurues* along the Magellan Strait, *cacaue* along the central western coast, cf., Viega Barros 2005 for linguistic and Martinic 1995 for historical evidence).

Modern paleobiological investigation reinforces this view of permeable frontiers and between group integration. Evidence of biological exchange between steppe hunters and southern canoe fishers, south and west of the Andes (Gonzalez-Jose 2003; Gonzalez-Jose et al. 2001) proves the inexistence of closed biological populations (see also Barrientos and Perez 2005 for the región North of Patagonia, and Morello et al. 2012 for the cross of Magellan Straits). Whether molecular markers (frequency of different haplogroups in the samples with the same geographical origin), or morphological and/or morphometric skull, variability can not be described in discrete units (or "types"), but as a continuum between more or less similar samples. Lalueza et al. (1997) argue that geographic distance (in a latitudinal sense) is the main factor that influenced the differentiation of the human groups from Tierra del Fuego and Patagonia. Recent studies at continental or subcontinental level in America prove that environment, diet and temperature, are the most important factor for understanding craniofacial and postcranial metric variation, explaining 50–80 % of morphological variation (Béguelin 2010; Beguelin and Barrientos 2006; Gonzalez et al. 2011; Fabra and DeMarchi 2011; Perez 2011). Therefore, the definition of biological "types" represents an incorrectly subjective assignment of affinities (Long and Kittles 2003; Gonzalez-José et al. 2008). Nevertheless, the alternating contraction and expansion processes of population (demographic and geographic), including events such as local extinction and recolonization of areas has important effects on the historical construction of cultural variability (Barrientos and Perez 2002).

According to our view, an isolation-by-distance model would predict that human groups reflected geographic separation in the pattern of their between-group distances. The eventual result would be a greater similarity between geographically proximal populations and increasing differences between groups that are further and further apart. The closest populations in space would have greater similarity in their biological characters than populations located further away. Biological differences observed among individuals who lived in Patagonia and those who lived in the rest of the subcontinent may be explained by a long history of divergence; current estimates range between 5,264 and 1,641 years of fissioning processes and isolation for the emergence of phenotypical differences from a single foundational population (García-Bour et al. 2003). Such a huge chronological range highlights the problems in the use of molecular clocks. Paleolinguistic research also suggests around 6,000 years for explaining the gap between *günuna a iajüch* and the languages from the *chon-tsoneka* family: *Chewache-iayich* (also called *teuschen*), *aónik'o ais* (also called *tehuelche*), *haush*, *selk'nam*, etc., assuming that both linguistic families come from a single foundational proto-language (Suárez 1970; Viega Barros 2005).

The historical trajectory can be tentatively reconstructed in the following terms. A relatively homogenous foundational population speaking a common language would have lived across the steppes to the east of the Andes with complex mobility and interaction patterns around 6,000 B.P., or probably before. Between 6,000 and 5,000 BP a noticeable reduction in the archaeological visibility in the northernmost part of this area (between 34° and 42° South) may be due to differences in mobility patterns, location of settlements or, more likely, the reduction in population density and population shrinkage due to migration processes and/or local extinction (Barrientos and Perez 2002, 2005; Boschín and Andrade 2011; Neme et al. 2011; Perez 2011). This would be the period were an original foundational identity and language *proto-gününa-chon* began to fission and evolve. The transition towards semiarid clima seems to have created the conditions for a later recolonization of the area by people of the same metapopulation, expanding from a few refugee areas, or members of a different metapopulation through processes or migration (Boschín and Andrade 2011). From 3,500–2,000 BP on, population expansion may have been affected also by the adoption of new technologies: bow and arrow and pottery. Barrientos and Perez (2002, 2005) suggest the existence of a strong biological relationship between groups of hunter-gatherers who occupied the Pampas and North Patagonia during the late Holocene (see also Béguelin et al. 2006; Cobos et al. 2012). While not yet established, it is possible that these groups would have configured a series of local populations belonging to a single metapopulation at a supraregional scale, experiencing contractions and expansions at different moments (Barrientos and Perez 2002; Barrientos et al. 2008; Prates 2008). The problem is that differentiated groups that may have existed were annihilated as a result of European colonization, and we have very poor information of linguistic diversity during early colonial times in Northern Patagonia. *Günuna a iajüch* is the only language of which we have some knowledge (Casamiquela 1983; Orden 2010), but there were many others.

Nothing of this population shrinkage in the middle Holocene and posterior phenomena of contraction and expansion is observable further south, between the rivers Chubut and Santa Cruz (42°–50° South lat.) (Belardi et al. 2010). Mena (1997) has suggested that between 6,000 B.P. and 3,000 B.P. this was a "macro-cultural region". It can be suggested that this is the original area where a *proto-chon* language differentiated from the northern *proto-gününe*. The individuality of those *proto-chon* speakers coincides with some observable differentiation in the archaeological record, like the general distribution of rock-art paintings and engravings and the specificities of lithic technology, in such a way that they can be used to distinguish this region (Orquera 1987; Fiore 2006; Gómez Otero et al. 2009; Cardillo 2011; Charlin and Borrero 2012). The ritual practice of cranial deformation is an additional evidence for the differentiation of northern Patagonia. South of 48° Latitude South, the frequency of this social behavior is very low. It has been suggested that the intentional deformations would have reflected the effort to achieve a predetermined cranial form, used as an indicator of group identity, demarcator of social or territorial boundaries, or as a trait that reinforces and maintains the networks of exchange between groups (Bernal et al. 2008; Perez et al. 2009). Around 2,500 B.P., speakers of a *proto-chon* language, already differentiated from their northern relatives (Suárez 1970; Viega Barros 2005), would have began to differentiate themselves, emerging new languages strongly related between them like *chewache-iayich, aónik'o ais, haush, selk'nam*, etc. It is interesting to note that the northernmost populations (not only speakers of *günuna a iajüch*, but also other unrelated linguistic groups) would be genetically and morphologically more similar to each other than to human populations further south, even though their supposed origin may have been different (Guichón 2002; Llop et al. 2002; Rothhammer and Llop 2004; García et al. 2006; Bernal et al. 2006; Pérez et al. 2007). This fact suggests a slower process of group fissioning in the south. This situation seems to agree, at least partially, with that suggested by Daniel Nettle for whom "the greater the problem of subsistence, the wider the social network necessary" (Nettle 1999). As a result, everything seems to indicate that "the greater the risk of not achieving the threshold of subsistence, the higher linguistic homogeneity exist in a geographical area of given size". However, this assumption should not lead us to uncritically affirm that the linguistic community was the basic social unit facing economic stress. Simply, contact with other groups must have been much more important in northern Patagonia than further south (Nettle 1999; Currie and Mace 2009). We may suggest that languages historically related as a result of the physical exchange of speakers are structurally and lexically more similar than those that were not connected and were also more geographically distant (Nichols 1997, 2008; Holman et al. 2007; Wichmann et al. 2008). The relationship that may exist between genetic distances, linguistic and "cultural" distances is the consequence of the fact that human populations (and therefore languages) "move" in a predictable way in some particular contexts. Therefore, the genetic distances between populations should be related in some way to the degree of statistical differentiation between the languages spoken by those people. The biological similarity among people and the existence of "cultural" differences in their motivations, behaviors and products should decrease as social interaction decreases as a result of an increase in geographical distance.

Both south and north, from the late Holocene onwards (ca. 1,000–800 B.P.), it has been suggested a reduction in mobility, the increase of population and a concomitant increase in the complexity of social interactions produced social instability, along with the emergence of a strong network of relationships between people culturally differentiated. In South Patagonia, this social and economic change has been related to the Medieval Climatic Anomaly, ca. 1,000 B.P. (Belardi and Goñi 2006; Goñi and Barrientos 2004; Goñi et al. 2007). Increasing aridity rates in this area would have caused the reduction of available fresh water sources, spatially constraining and concentrating animal movements and human settlements, and leading to greater social specialization in the use of physical space. This could have created conditions for habitat fragmentation, a local increase in population density and increased spatial coherence. The opening of social exchange networks would have compensated for the reduced mobility of residence patterns and the nucleation of human settlement. For instance, the radius of movement of rocks and raw materials would have extended to 800 km (Gómez Otero 2007). At this level of differentiation, kinship and political alliance constituted the only mechanisms for fixing the limits of social groups, which differed in size, language, culture, social structure, and probably also in the nature of their predominant economic activities. Archaeologically, the high rates of burial area reuse would suggest that human groups were increasingly fixed to specific territories (Gómez Otero 2007; Borrero et al. 2008). The concentration of rock-art on the Stroebel Plateau would suggest the aggregation of different groups at specific places.

Although such a climatic phenomenon would have had different effects at different places (Favier Dubois 2004), a similar transition has been suggested for North Patagonia. There, strongly differentiated human groups would have shared the same process of economic and social intensification consolidating complex social-political networks that favored the movement of goods, people and ideas in a very large social space (Mazzanti 2006; Luna 2008). Precisely in this period, there is clear evidence of a more intensive occupation in some areas, and a significant variability in mortuary practices. By 1,000 BP, there was a transition to the current weather conditions and retraction of the main animal resources to the West and South of the Interserrana area, what probably implied the redistribution of the existing population and/or the expansion of another population(s) from northeastern Patagonia. In the latter part of the late Holocene (ca. 1,000–400 years BP), there is growing evidence of a population expansion from the lower basin of the Colorado and Negro Rivers and Atlantic coastlines, to the plains on both sides of the Ventania Sierra. The potential competition between the local population and the new immigrants would have favored the latter, which reached a dispersion range that included the aforementioned areas and at least part of the areas Tandilia Interserrana and Serrana (Barrientos and Perez 2002, 2005; Béguelin et al. 2006). Craniological studies by Barrientos and Pérez (2004) suggest the presence of expanding populations from northeastern Patagonia to Southeast and southern Pampas. Furthermore, the bioarchaeological record from the south-central La Pampa province seems to reflect two different populations in just north of Northern Patagonia (Berón 2005). Finally, Gonzalez-Jose (2003) has recognized morphological affinity between skull samples of the foothills of northern Patagonia, the Black River valley and northeast and the Pampas of Buenos Aires, probably due to interbreeding.

The later progressive homogenization of languages and cultures across eastern continental Patagonia was probably caused by an increase in the frequency and intensity of long-distance exchange mechanisms (Lazzari and Lenton 2000; Mandrini 1991; Palermo 2000; Villar and Jiménez 2003; Nacuzzi 2007, 2008; Carlón 2010). Archaeologically, this process can be inferred from the increase in population, more sedentary occupations, symbolic manifestations (rock art), technological innovation (ceramics and specialized instruments), formal burial areas, foreign exchanges, etc). The even greater complexity, intensity, and frequency of social interaction between groups determined the transformation of traditional means of social reproduction and political order. Mechanisms for collective decision-making began an ever-increasing hierarchization process, concomitant with the increased size and more diverse composition of human groups. Social relations of production began to acquire some characteristics related with domination. To sum up, we must avoid the traditional mistake made by the first European travelers in Patagonia and the early ethnographers who described indigenous groups as if they were Old World nations. According to all evidence, ethnic, linguistic, cultural, economic, territorial frontiers were extremely permeable, suggesting a considerable degree of population mixture. Consequently, the apparent cultural unity recorded by modern ethnographers was just a phase in the changing nature of social exchange, and not a fixed cultural trait since the origin of those populations (Boschín 2001, 2002).

10.4 An Agent-Based Simulation Model for Understanding the Emergence of Patagonian Ethnicity and Territoriality

We have built a computer simulation (see Fig. 10.1) to explore how ethnogenesis and related process of territorialization could have occurred in the prehistoric past of Patagonia. The current implementation is a further development of previous, preliminary attempts, partially published in Barceló et al. (2013a,b), Barceló and Del Castillo (2012), Del Castillo (2012), and Del Castillo et al. (in press). The new computer program has some important advances in the way positive interaction has been modeled, and in the modelling of the mechanisms of social reproduction. The number of free parameters has been reduced and some important non-linearities have been taken into account. Programming code is implemented in Netlogo (Wilensky 1999) and provisionally available from http://www.openabm.org/model/4063. A full description of the algorithm appears in Barceló et al. (2013b).

In the model, agents simulate "families" or households, defined in the following terms:

- *Labor* (l_i), a Poisson distributed parameter counting the aggregated quantity of labor from all family members).
- *Cultural identity*, a vector of 10 dimensions; each component is an ordinal number from 1—not important—to 6—very important. Such dimensions are weighted according to a fixed vector).
- *Technology* (β_i), a Gaussian distributed parameter for each agent affecting the efficiency of labor when obtaining resources.
- *Energy-conservation factor* (d_i) calculated as $\beta_i/2$ and expressing the efficiency of storing and preservation methods: the part of acquired energy that can be stored and transferred to the next time-step.
- *Survival threshold* ($\bar{e}_i$): Given that the survival of agents depends on the amount of energy acquired and transformed from the environment, and the number of members the household has (expressed in labor units), a survival threshold should be calculated in terms of the quantity of calories an agent (representing a group of individuals) needs to be able to live a season long. In the simulation the household size is equivalent to labor. Assuming an individual needs an average of 730 kilocalories per year (2,000 calories per day; based on estimations by the Institute of Medicine, 2002), and one time step (cycle or "tick") in the simulation roughly represents what an agent is able to do and move in six months, $\bar{e}_i = (730 \times$ the number of labor units at this agent)/2.

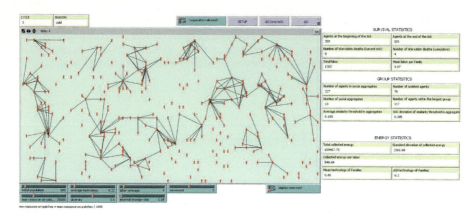

Fig. 10.1 A screenshot of the interface up front

The model's diversity index expresses the amount of variation between agents for reasons characteristic of the agent, and not of global demographic factors. We have assumed it is a global Gaussian parameter measuring the standard deviation of productive instruments (β_i) and storing means (d_i). We do not have precise estimates (but see Binford 2001), so we have fixed this parameter with a medium value (diversity = 0.5).

Physical space is modeled as a 40×80 grid, and it contains 3,200 environmental cells or "patches". We assume that each grid cell is a surrogate of a 100×100 km geographically homogenous area, interpreted as the total extension a virtual household can explore during a season of six months in its search for resources and people. Each path has a number of *resources* (r_i), a random distributed parameter, measured in kilocalories, diminishing at odd cycles ("cold" season) and reproducing the original value at even cycles ("hot" season) to reproduce seasonality. Resources at each patch have also a *difficulty* level (h_i) (another random distributed parameter). It counts the difficultness of resource acquisition (the more mobile the resource—animals—and the less abundant, the more labor or more technology is needed to obtain resources up to survival threshold. The availability and abundance of resources are assumed to variate randomly through the landscape; therefore we have used a uniform distribution of values between a minimum and a maximum value. From a theoretical minimum value of 100 kilocalories, we have explored different intervals: from 100 to 15,000 kilocalories (the "poor" world hypothesis), from 100 to 20,000 kilocalories, from 100 to 25,000 kilocalories, from 100 to 40,000 kilocalories, from 100 to 50,000 kilocalories (the "rich" world hypothesis). Such configuration intends to simulate the way edible resources were distributed in the Patagonian past. The main source of food was the locally evolved camelid *lama guanicoe* ("guanaco") and although very mobile, numerous herds dominated the landscape (L'Hereux 2006; Gómez Otero 2007; Papp 2002; Prates 2009; Politis et al. 2011; Rivals et al. 2013). The consequence is the existence of a source of subsistence that can be occasionally and locally abundant but spatially and temporally variant and relatively unpredictable (Soriano et al. 1983; Paruelo et al. 1998; Schulze et al. 1996; Borrero et al. 2008; Mazia et al. 2012). The model implements a simplified seasonality: a hot season in which natural resources are initialized to its maximum value, and a cold and dry season in which resources do not regenerate naturally, and the amount of resources available in each cell is equal to the half of what was generated at the hot season minus what the agent extracted at the previous time-step. In any case, our simulated environment does not pretend to reproduce Patagonian ecology. It is obvious that landscape differences and topographic barriers would have affected hunter-gatherers subsistence and mobility. Instead, we want to investigate what could have happened if geography played no role in social dynamics.

The way in which Patagonian hunter-gatherers defined, conceived and behaved regarding resources and subsistence did not meet universal standards, but was mediated by a complex and unique system of practices and beliefs, influenced by the characteristics of the resource itself and the general environment for energy needs, and the social, ideational and historical trajectory of people (Prates 2009). Therefore, we have not considered the individuality of each resource, but the human results of the activity. Energy is obtained by agent i by means of labor ($l_i(t)$) with the contribution of its own technology, whose efficiency is estimated as $\beta_i(t)$. Both factors act upon the difficulty of acquiring and transforming resources, in such a way that:

$$f_i(t) = \frac{1}{1 + \frac{1}{h_i(t) \cdot l_i(t)^{\beta_i(t)}}} \qquad (10.1)$$

$f_i(t)$ measures the ability to obtain resources according to each agent's individual ability. Its maximum value is 1, indicating the amount of work available (l_i) and the effectiveness of current technology β_i to compensate the local difficulty (h_i) of obtaining the resources existing at that place. When the value of $f_i(t)$ is less than 1 (but greater than 0), we can deduce that the working capacity and technology available only allow obtaining a proportion of the available resources. We are not taking into account the precise energetic performance of each resource, vegetal or animal, but the probability of attaining full survival with an undetermined series of resources obtained locally.

We assume that the higher the technological level, if the amount of labor does not vary and local resources remain stable, the less cooperation and lesser chances of cultural diversity. That means, that hunter-gatherer groups with poor technology based on worked stones and transformed wood will manifest higher cultural homogeneity than groups with a technology that allows them to transform into subsistence all existing local resources. The *technology* parameter may range from 0.01 to 2. High efficiency indicates that all local resources can be managed independently of its difficulty of acquisition given the extreme performance of available technology. Low values are characteristic of human groups with hardly evolved instruments, in such a way that only a part of locally available resources are effectively managed. The efficiency of food preservation techniques is another technological factor, related with the overall level of development of means of production. In the experiments we report here, we have fixed parameters related with technology and efficiency using data from our own research in Patagonia (Barceló et al. 2009, 2011; Del Castillo 2012): average-technology $= 0.22$ (low development); standard deviation (diversity among simulated households $= 0.5$); storing capability $= 0.11$ (very low). In the absence of efficient hunting equipment beyond "boleadoras" and spears (bow-and-arrow was a relatively late instrument in Patagonian archaeological record, and hardly adapted to the capture of local game).

Virtual households can be involved in two kinds of economic activities: gathering, which is an individual task, and hunting, which is only possible as a collective task. Ethnographical sources make manifest the difficulties of hunting guanacos, and the need to ask for the help of many people to encircle the game and be able to kill enough prey (Fitz-Roy 1932 [1833–1839]; de Orbigny 2002 [1833]; Cox 1999 [1862–1863]; Claraz 1988 [1865–1866]; Musters 1964 [1872–1873]; Spegazzini 1884). At the beginning of twentieth century, a witness described:

> Leaving early in the morning they rode out into the camp. They had already ascertained where several pregnant guanaco were feeding. The riders lined up in a huge, loosely knit circle about them, unnoticed, and at an appointed time all rode in towards the center. The game ran, only to meet other riders, ran from them, to meet others on the shrinking circle. If any broke through, a rider balled it, jumped quickly from his horse and killed it, mounted and was back in place in no time. Lions, ostrich, deer, and guanaco shared the same fate. The trapped animals fought to escape when the ring drew close about them, and the Indians, in a sort of ecstasy, caught and killed as many as they could. If there were riders enough, and good horses under them, few would escape, and at last the center would be a mass of dead animals or struggling live ones, killed or entangled by boleadoras. (Childs 1936, p. 160f)

In our simulation, "hunters" need the contribution of other hunters in the neighborhood. The aggregated productivity $[\Delta f_i(t)]$ of an agent member of a group $G_i(t)$ is calculated as:

$$\Delta f_i(t) = \cfrac{1}{1 + \cfrac{1}{\left[h_i(t) \cdot \left(\sum_{j \epsilon G_i(t)} l_j(t)^{\delta \beta_j(t)}\right)^{\theta_i(t)}\right]}} \tag{10.2}$$

where $G_i(t)$ is the total amount of labor the group of agents that cooperate with agent i and $\delta \beta_j(t)$ the maximum technology within the group. There is an additional parameter modifying the total effect of aggregated labor at the social aggregate $(\theta_i(t))$, illustrating the idea that cooperation is less needed when there are plenty of resources.

Agents cannot move to an occupied patch, so they never share their resources. What they share is labor, and not the products of that labor. Sharing labor and technology is a way to increase the chances of survival when the productivity of the patch (quantity of resources modulated by labor and technological efficiency) is below the survival threshold. By doing so, agent i receives cooperation in form of labor. There is no obligation to "return the favor": only the helped agent receives help when its similarity threshold between the helper and the helped is low enough so that the helper "can afford" to help. There is no compensation for the excess of labor exchanged, or calculation of differential costs. This is not a limitation of the model, but a phenomenon that is well understood in the ethnography of hunter-gatherers. Given that labor attains its limit when survival is assured, there is no surplus. Consequently, there is always a remanent of "unused" labor. When hunter-gatherers aggregate, all members identify themselves as members of the same group, and all labor is put in common. We assume agents in the simulation do not use the fiskean logic of "Equity matching" but a form of "community sharing" (Fiske 1991). Ethnographic sources suggest that the decision to cooperate or refuse cooperation was far more complex in Patagonia than the simplified approach adopted in PSP 1.5 (Martinic 1995; Papp 2002). There are some common aspects, however: it seems that cooperation within the kinship network was far more frequent than with strangers, and that kinship ties were constantly negotiated even without marriage exchange (Musters 1964 [1872–1873]; Fernández Garay and Hernandez 2004). Our algorithm follows such kind of limited and changing parochialism.

Cooperation at work and the consequences of its restriction are at the core of the simulation. Agents decide to cooperate and work together when at least one of them needs the help of others to obtain enough resources for survival and there is someone in the neighborhood able to cooperate given the relative similarity of social values. That is to say, to decide if an agent cooperates with another, we imagine each one observing the immediate neighborhood and evaluating their respective identities to know if they are "sufficiently" common. Each agent has its own organized list of meanings, values and beliefs (*identity*), inherited at birth, learnt within the evolving group, modified all along the life of the agent and transmitted to the new generation. Agents rank the relevance of each value-dimension according to a fixed weight vector. Thus, they capture the agent values without explicitly identifying values as the topic of investigation. The simulation asks about similarity to another agent with particular goals and beliefs (values) rather than similarity to another agent with particular traits. Consequently, instead of *assuming* that agents have common identity traits based on membership to an already existing "ethnic" group, agents need to be queried as to the extent to which they "believe" they are similar to those of others in the neighborhood, and queried as to whether the outcomes of those values are perceived to be similar.

When cooperating and sharing labor capabilities, information and knowledge flow between agents. Therefore, the most effective technology proves its advantages and begins to be adopted by members of the same group. Technology should be updated in such a way that the next tick will increase its efficiency towards the level of the most efficient within the group. This is a process of convergence and not of imitation. An additional source of technological evolution is implemented in form of an *internal change rate* (hereafter ICR). This is a random value (from 0 to 1, usually very small) defined as a random factor that expresses the likelihood of internal change (invention, catastrophe, sudden change) affecting values and technology. The higher this value, the more important internal changes in the virtual "family", expressing the probability that each agent changes independently of the other agents.

Given the evidence of the Patagonian archaeological record and its 7,000 years of technological continuity, we have assumed a very low likelihood of internal change (0.05), according to the archaeological evidence of slow technical, linguistic and cultural transformations in Patagonia (Barceló et al. 2009; Del Castillo 2012).

With probability equal to ICR, the agents adopt a new technology value, whose average is calculated on the basis of the global parameters: average-technology and diversity. Technological involution has been an exception historically, and we do not take this into account in the model. Although technological change is mostly "rational" at the scale of the individual taking the decision to change, from an external perspective, such decisions at the local level may appear as internal shocks perturbing the apparent linearity of a given trajectory. Therefore, although technical evolution is not random at the level of each agent, it can be modeled as random at the level of the population.

As a result of interaction and information flow, cultural consensus emerges by combining the identities and values of interacting agents in an emergent group. Therefore, once the agent gets enough resource for its own survival (with or without the help from others), the identity vector used to define the possibilities of cooperation is updated towards the statistical mode of the group identity. With a fixed probability level (95 %) each agent copies the statistical mode of identities within the group. There is an additional source of identity change, also implemented in the form of the same ICR we have considered before. With every tick, and with a fixed probability level determined by the opposite of the identity weight vector, the identity vector mutates. In this way, we assume that the most "universal values" are the least prone to internal change (although they may change, but with lower probabilities). The most frequent internal changes appear in the less "universal" dimensions, which are also the less relevant to build cultural consensus. Therefore, what future generations arising in the aggregation will inherit is not the old identity, but the new commonality. We assume that the higher the cooperation between people, and the higher the cultural consensus among them, the higher the probability that reproductive couples will be formed within the group. The idea is that once the new social aggregation has emerged and survival of agents has been assured, hybridization mechanisms begin to act because inherited identities (ethnicity) should be modified to maintain the newly built consensus.

Because hunting is more productive, there are increasing returns to collective involvement. Survival is also affected by *diminishing* marginal returns relative to the social and technological impossibility to regenerate resources, and the need to wait for a minimum of one year for its natural regeneration. Agents lose one of its members (a labor unit) each time the total acquired energy is below the survival threshold. In the same way, every 30 ticks, a new member is born, and will live until the total acquired energy is below the survival threshold. In this way, we have implemented a determinist population growth mechanism opposed to stochastic mortality. When survival is possible and the number of members in an agent (expressed in labor units) is greater than 10, the current agent reproduces and gives birth to a new agent, with half the parent labor, the same technology and the same identity.

Agent actions are oriented to foraging and food gaining through mobility across a territory, conditioned by available technology and agent density, and the establishment of cooperation between agents when direct survival is not possible. However, what they have acquired as subsistence has a short temporal duration, and given the low degree of storing technology, agents should begin the process anew at the beginning of each time-step. In the model, the availability of resources is fixed as a global probability parameter ("rich world", "poor world"). Each agent has the possibility to move camp/settlement location and interact with other agents in order to decide whether to cooperate or not in survival or in reproduction. The agent has the goal of survival at least after T simulation steps. To do that the aim of each individual is to optimize the probability of survival and reproduction by gaining enough food (energy reserves) to meet a threshold of energy necessary for successful reproduction.

Agents should take the decision whether to move to another place with more resources, but where positive interaction with others may not be possible. According to that decision, each agent may remain in place interacting with the same agents it interacted with at a previous time step, or it can move to another patch. Agents move randomly because they can follow any direction within a restricted neighborhood. When moving, agents first determine who it can cooperate with from the group of agents in place (my_group). The process identify-agents is based on a calculation of the number of common identity traits perceived among agents within a neighborhood. An agent does not have information about all the agents in the world, but only those within a reasonable geographic distance (my_neighborhood). The extension of such a neighborhood simulates the precise territory agents arrive to know by themselves or by means of communication flows from linked agents. The size of the neighborhood changes as a consequence of the displacement of the center of the neighborhood, maintaining the same radius (a limit connected with the low efficiency and efficacy of transport technology). In this way, the model has an emphasis on local dynamics and bounded rationality. Whether cultural consensus is high enough, agents are listed into the newly emerging group, and the program characterizes such a group with a distinct color. Once within an aggregate (my_group), an agent's subsistence output can be enhanced adding to the agent's capacity to work, and the capacities of other agents within the group.

Identity traits have been modeled as adaptive behaviors, because in some sense they act to increase a measure of the virtual household success at meeting some goal (survival). In so saying, we assume some degree of "utility" for agents' identities: if they change and negotiate their identities they can obtain higher probabilities for success. Consequently, we assume each agent's goal is to maximize survival probabilities through increasing the probability to hunt with the help of labor from other agents. However, the only "rational" decision executed by an agent is the decision about whether it cooperates with a neighbor or not. Such a decision is carried out by computing the resemblance of identities, and using a changing decision rule for determining the degree of similarity in terms of the circumstantial needs and expectations from collective hunting. That is to say, our virtual Patagonian

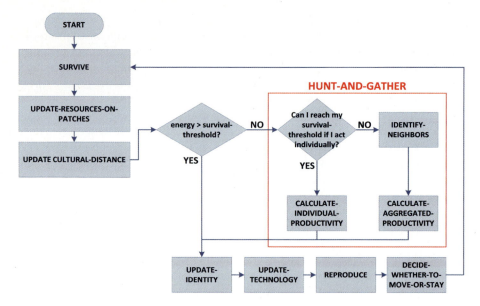

Fig. 10.2 The model's flow-chart

households are able to change the value of their tolerance to cultural and social diversity given the actual needs to enhance probabilities for survival when resources are spatially or temporally scarce.

Figure 10.2 summarizes how the model runs. At start up, agents are placed randomly in the world. Each agent should occupy a single patch, and no two agents are allowed to occupy the same patch. If their energy level is below their survival threshold they look for resources (*hunt*-and-*gather*), constrained by the amount of work available at a single household (*labor*) and the current value of their technology. If acquired resources are not enough, they look for neighbors to cooperate with. If no one cooperates, and resources remain below survival threshold, the agent dies.

10.5   Running the Model: Preliminary Results

The current version of the model differs from Patagonian ethnoarchaeological data in some important ways.

We have implemented a single, homogenous founding population, although current paleobiological research seems to conclude the likelihood of a minimum of two or even three well differentiated founding populations (Gonzalez-José et al. 2008; Lewis et al. 2007; Pérez et al. 2007; Rothhammer and Llop 2004; Bodner et al. 2012). Miotti and Salemme (2003) have suggested that early settlers would have belonged to a Patagonian megapopulation that would have split in northern South America, moving independently on both sides of the axis of the Andes, which would have acted as biogeographic filter. This would have led to processes of colonization and expansion-retraction differing between the two slopes. However, the hypothetical difference in founding biological populations is still under discussion and there is no hard evidence about it. Our simulation intends to explore what could have occurred in the case of a single population as it first colonized a previously unoccupied landscape and the increasing differences between groups emerged as households and grew further apart in their constant movement in the quest for resources. How do processes of convergence and divergence occur between groups of hunter-gatherers over the long-term?

The simulated environment has nothing to do with environmental conditions during the Holocene in Patagonia. We have not modelled a "virtual Patagonia", and we have explicitly avoided the representation of geographical details. We know that in prehistoric Patagonia, human groups aggregated where resources were more abundant, temporally frequent and easy to get, but we doubt that the environment was the only cause. What would have happened if the environment had no influence on spatial aggregation? We have imagined a cold and dry plain without any topographic features, where resources randomly varied from very scarce to very abundant. We have experimented with all possible scenarios, beginning with a very "poor" environment, and finishing with the "richest" imaginable one. If resources in the environment are scarce (below 15,000 kilocalories per patch), a small population (estimated at 300 "families" with an average of four members in each; based on estimations published in Papp 2002), with hardly efficient technology (both for producing and for storing), would never survive on their own (without any kind of cooperation with neighbors) beyond 100 simulated years. In this simulated scenario, the wealth of resources clearly influences survival in a linear way ($r^2 = 0.688$). However, when virtual households with similar identities exchange surplus labor and share the most efficient technology, mortality clearly reduces, and the influence of resources was clearly non-linear ($r^2 = 0.365$). In other words, when our simulated Patagonian hunter-gatherers interacted and worked together, the probability of their survival was higher than if they had worked only on their own.

Technological efficiency experienced changes and evolution, both in prehistoric Patagonia as in our simulation. Here computational results coincide with archaeological data: there is evidence of small but continuous changes, interpreted as local advances not related with interaction, but also a gradual convergence towards the most efficient, when innovations diffused. Figure 10.3 shows how in a cooperation scenario, average-technology quickly evolves towards more efficient values as a result of innovation-diffusion through conspicuous imitation and borrowing. The diagram shows interpolated curves, that although in their first part seems to have a lesser than average model, they correctly predict the temporal trajectories. The difference of means has been proved to be statistically relevant.

Cardillo (2011) has shown how both environment and geography account for a statistically significant part of the lithic technology variation. The archaeological pattern is much more detailed than in our model, suggesting a latitudinal gradient

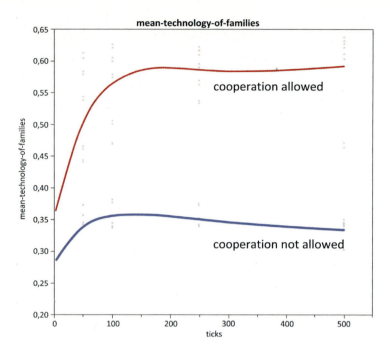

Fig. 10.3 The temporal evolution of average-technology after 500 runs (simulated 250 years). We have here averaged the different wealth scenarios: *grey dots* represent original data from which both curves have been interpolated. Graph computed using JMP 10 software (SAS, Inc.)

in diversity that might be explained as the result of restrictions of information borrowing within a culturally homogenous population (parochialism) as well as of selective mechanisms related to energy acquisition (see also Gómez Otero et al. 2009; Charlin and González-José 2012).

As a result of economic interaction, virtual households aggregate in space, configuring what we can consider social networks of cooperation. The model does not predict the formation of discrete groups with clearly defined borders and frontiers, but the emergence of changing networks of social relationships, with different possible topologies: in some contexts, closed groups may emerge, but when the intensity of interaction varied, or the circumstances in which the interaction took place were different, the nature of the social aggregation was also different, allowing the dissolution of any previously differentiated group into an undefined homology of social activities.

We have suggested that commonalities in needs, motivations, goals, actions, operations, signs, tools, norms, cooperative ties, and in division of labor schemas are the consequences of the way some social agents interacted, aggregated in space and time as a consequence of some of these interactions, and reproduced the basis of such an aggregation. Our model suggests that in prehistoric Patagonia, social aggregation and network formation may have been more frequent in the cold season,

Table 10.1 Parameters and results for the different scenarios shown in Fig. 10.4

Name	Scen. 1	Scen. 2	Scen. 3	Scen. 4	Scen. 5
Max. resources	15,000	20,000	25,000	40,000	50,000
Initial agents (founding households)	300	300	300	300	300
Total number of agents after 250 years	40	222	351	968	1,144
Number of social aggregates	3	14	12	2	1
Number of agents in social aggregates	8	116	220	709	868
Number of isolated agents	32	106	131	259	276
Number of agents within the largest group	3	29	45	690	868

given a higher frequency of aggregation. During the hot season the benefits of cooperation are less obvious and therefore the probability of any form of restricted territoriality is significantly lower. According to the Analysis of Variance (ANOVA) test on 100 examples of each wealth scenario, the probability of equal average number of social aggregates in the hot and cold season is less than 0.001 in all scenarios. Exactly the same is true for the size of the network—the number of agents in social aggregates. This result is compatible with ethnoarchaeological evidence in Patagonia: with bigger campsites in cold seasons and a general dispersal of households during the hot season (Moreno and Izeta 1999; Boschín and Del Castillo 2005).

Simulation results (refer to Table 10.1) correlate with J. Gómez-Otero's reflection on the need for "places of concentration and distribution" (Gómez Otero 2007): She cannot consider Patagonian human groups randomly wandering on foot, at any time of year, to find someone with whom to cooperate. No hunter-gatherer would invest so much energy in search times if there was no assurance for success in meeting and obtaining searched resources. Our simulation predicts that very few groups will keep moving again and again. Rather, some kind of "good-enough" scenario is found where groups stay in the neighborhood of other groups, keeping the connections among them.

Are such networks an initial form of ethnogenesis? We stressed at the beginning of this paper that *the lesser the intensity and frequency of inter-group relationships, the greater the differences in ways of speaking and other cultural features manifested by groups.* The same can be said in terms of network embeddedness. Network embeddedness means that everybody does not interact equally with everybody else, but is constrained by needs (expected benefits), geographical neighborhood and prior cultural consensus (common history). Agents within the network interact among themselves more often than with others out of the network, which means that a subset of the population may be excluded from positive interaction and hence the process of similar identity negotiation and innovation diffusion. How intense is the resulting segregation in the explored virtual scenarios? We have measured it in threesteps: fractionalization, generalized resemblance and demographic polarity.

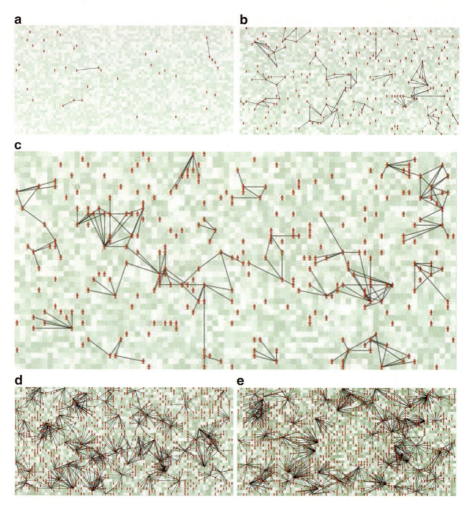

Fig. 10.4 Different scenarios of virtual Patagonia, variating the maximum resources at patch. Links visualize agents that cooperated (exchanged labor) at the current tick. Screen-shots of the simulation after 500 time steps (ca. 250 years). Parameter settings for all scenarios are given in Table 10.1. (**a**) Scenario 1. (**b**) Scenario 2. (**c**) Scenario 3. (**d**) Scenario 4. (**e**) Scenario 5

A traditional measure of social fractionalization can be calculated by dividing the population into ethnic groups, calculating each group's share of the population, summing the squared shares, and subtracting the sum from one. Such a measure was calculated by Taylor and Hudson (1972) as a decreasing transformation of the Herfindahl concentration index applied to population shares. In particular the index takes the form of

$$ELF_j = 1 - \sum_{i=1}^{I_j} \left(\frac{n_{ij}}{N_j} \right)^2 = 1 - \sum_{i=1}^{I_j} s_{ij}^2 \qquad (10.3)$$

where n_{ij} is the number of people that belong to ethnolinguistic group i in country j. N_j is the size of the population in country j and I_j is the total number of ethnic groups in country j. This formula requires the groups to be mutually exclusive (i.e., if an agent is in aggregate 1, then it is not in aggregates 2-n) and exhaustive. Given mutual exclusiveness and exhaustiveness, this index measures the probability that two randomly chosen individuals from a country's population belong to different groups. The measure scores zero where in a perfectly homogenous population (i.e. all individuals belong to the same group) and reaches its theoretical maximum value of 1 where an infinite population is divided into infinite groups of one member (Alesina et al. 2003).

   In our case, we have simplified calculations which do not take into account isolated agents. In fact, each isolated agent would have constituted a differentiated group, so actual results should offer higher fractionalization indexes that those provisionally calculated here (see ELF score in Table 10.2).

   Fractionalization increases when the number of small groups increases. In our case, the probability that two randomly drawn individuals from the population belong to two different groups increase when resources are low and survival may be at risk. The higher the value, the higher horizontal inequality in the total population. These results are very interesting for understand the consequences of the Medieval Climatic Anomaly, ca. 1,000 B.P., in some Patagonian areas. The reduction of available fresh water sources would have spatially constrained and concentrated resources and human groups, and created conditions for residential fragmentation. Our results clearly show that when the simulated world is comparatively poor (maximum resources less than 30,000 kilocalories for a complete season), as during the Medieval Climatic Anomaly, fractionalization scores are higher than in the case where resources are abundant and frequent. Following Vigdor (2002), estimated fragmentation effects can be interpreted also as the weighted-average of within-group affinity in the population. That is to say, a high value of fragmentation when resources were scarce and concentrated can be explained as the probability of an

Table 10.2 Further calculation results

	Scen. 1	Scen. 2	Scen. 3	Scen. 4	Scen. 5
ELF score	0.9863	0.92	0.97	0.423	0.431
Minimum Euclidean distance between households	0	0	0	0	0
Maximum Euclidean distance between households	7.0	7.5	7.34	9.94	6.70
$G(S_N)$	4.96	0.86	0.06	0.115	0.144
RQ	0.0511	0.14	0.09	0.56	0.56

individual's willingness to spend on available resources given the degree of affinity within its constrained neighborhood. The probabilities of successful economic interaction vary depending on how many members of the community share the same identity of that individual. It is important to take into account, however, that our results are not linked to a specific moment in Patagonian historical trajectory. To the extent that social aggregates are constantly changing, especially between the hot and cold seasons, ELF scores never remain constant. Calculated values only refer to a specific state of the simulation (500 "ticks", or 250 simulated years).

It is usual to explain the effects of the Medieval Climatic Anomaly in Patagonia in terms of the potential competition between the spatially differentiated populations, with the emergence of "territoriality". Different authors (Belardi and Goñi 2006; Goñi and Barrientos 2004; Goñi et al. 2007; Gómez Otero 2007) suggest that during the peak of greatest aridity, the presence of water in the environment may have become circumscribed to specific loci (e.g., relict lake and permanent watercourses) that would have had the potential to act as hubs for population aggregates. Human groups reduced their residential mobility, so that settlements were confined to locations with availability of critical resources (water, wood) and good condition (repair, mild winters). Parallel to this reduction in residential mobility, the ranges for logistic action would have expanded and extended. Among the consequences of these circumstances, a decrease in population density at a regional scale has been suggested, whereas density increased locally. Our results are congruent with these hypotheses. Our results also seem to coincide with those of J. Gómez-Otero (2007) which has suggested a "gradual" population growth at this period, with very localized moments of stress and competitive concurrence.

In our simulation, the index of fractionalization is just a measure of hetero-geneity; such measure conveys no information about the depth of the divisions that separate members of one group from another, which is a necessary factor for inferring *social tension* (Fearon 2003; Posner 2004; Chandra and Wilkinson 2008; Brown and Langer 2010; Chakravarty and Maharaj 2011). The ELF index can at best be seen as a measure of cultural diversity but not a proxy for the effect of diversity as a whole. We may arrive at the depth of the "difference" in terms of the non-normalized Euclidean distances (see Table 10.2) between *cultural identity* vectors (see definition on page 227; note that in the rest of this chapter, we usually omit *cultural* and just talk of *identity* or *identity vector*). In our simulated world, at time-step 0, this value is 0 because the founding population is supposed to be homogenous. Two hundred and fifty simulated years after, the differences have clearly increased: although some households maintain their similarity (Distance = 0), many others have augmented their differences (maximum measured Euclidean distance is 9.94).

The reference value of Maximum Euclidean distance between identity vectors in our case is 18.97, which results when identity vectors (ranging from 0 to 6, as defined earlier) are totally different:

$$0\ 0\ 0\ 0\ 0\ 0\ 0\ 0\ 0\ 0$$
$$6\ 6\ 6\ 6\ 6\ 6\ 6\ 6\ 6\ 6$$

Bossert et al. (2011) and Kolo (2012) have introduced a more flexible version of the ELF, the generalized ethno-linguistic fractionalization index. Based on the specific characteristics, a mutual similarity matrix between individuals takes the distance between them into account. Hereby the groups emerge 'endogenously' from the matrix. The similarity value between two individuals i and j for all $i, j \in 1, \ldots, N$ is given through s_{ij}. For a society with N individuals, all s_{ij} are contained in a $N \times N$ matrix, labelled similarity matrix S_N, which is the main building block of this measure. Based on this matrix, the corresponding generalized resemblance value for a population with N individuals is given through:

$$G(S_N) = 1 - \frac{1}{N^2} \sum_{i=1}^{N} \sum_{j=1}^{N} s_{ij} \qquad (10.4)$$

In calculating $G(S_N)$, each individual counts in two capacities. Through its membership in its own group, an individual contributes to the population share of the group. In addition, there is a secondary contribution via the similarities to individuals of other groups. In our case, and considering the state of the agents' identity similarity after 500 ticks, we get the results given under $G(S_N)$ in Table 10.2.

Those results should be interpreted as the expected dissimilarity (in Euclidean distance terms) between two randomly drawn individuals. In our case, the poorer the world, the higher the expected dissimilarity. When the world seems rich enough and fractionalization is less conspicuous, expected similarity is far greater. These results seem to be concordant with the process of cultural hybridization at the end of the Holocene. What was fractionalized when resources were scarce and concentrated became homogenized when technology increased suddenly its efficiency (imported colonial items, horse domestication) and resources increased by foreign factors (horse domestication, acquisition of colonial cattle and new technologies) (Mandrini 1991; Mandrini 1992; Palermo 2000; Villar and Jiménez 2003; Nacuzzi 2007, 2008; Carlón 2010). The idea of the "tehuelche complex" (Escalada 1949; Casamiquela 1965; Martinic 1995; Papp 2002), an integration and hybridization of a plurality of previous identities into a new syncretism would also relate with such results.

Generalized resemblance does not solve our problem about the emergence of segregation and territoriality when group fractionalization increases. Obviously, if dissimilarity is great and fractionalization is intense, the probability of competition should be higher. But the number of groups and the degree of difference on their own are not enough to conclude social tension and violence. "Polarization" is needed to transform difference into competition. Theoretically, polarization should be calculated in terms of the "distance" between two groups, i and j, corrected by the sizes of each group in proportion to the total population (Esteban and Ray 1994; Duclos et al. 2004). The assumption behind this alternative measure is that whilst the generalized fractionalization matrix rightly attributes a low chance of ethnic conflict to an homogenous population, highly fractionalized populations are not conflictual as no group has the "critical mass" necessary for conflict. Conflict will be more likely the more a population is polarized into two large groups, well beyond a specific critical mass. Montalvo and Reynal-Querol (2002, 2005; Chakravarty and Maharaj 2011) have developed an index of demographic polarization

$$RQ = 4\sum_{i=1}^{k}\sum_{j\neq i} p_i^2 p_j = 4\sum_{i=1}^{k} p_i^2(1 - p_i) = 1 - \sum_{i=1}^{k}\left(\frac{0.5 - p_i}{0.5}\right)^2 p_i \qquad (10.5)$$

p_i in the equation is the proportion of people who belongs to ethnic group i. RQ employs a weighted sum of population shares. The weights employed in RQ capture the deviation of each group from the maximum polarization share 1/2 as a proportion of 1/2. Analogously to the index of fractionalization, underlying the formula for RQ is the implicit assumption that any two groups are either completely similar or completely dissimilar, and thus the weights depend on population shares only. This index tends towards zero for very homogeneous and non-conflictive populations, i.e., with only one relevant group. However, with increasing group numbers, ELF and RQ show clearly different results. While ELF is an increasing function of the number of groups, RQ reaches its maximum with two equally sized groups (i.e. $i = 2$, $p_1 = 0.5$, $p_2 = 0.5$) and decreases afterwards. It is the same to say that social heterogeneity and social conflict is not one and the same. Initially, one could think that the increase in diversity increases the likelihood of social conflicts. However, this does not have to be the case. In fact, many researchers agree that the increase in ethnic heterogeneity initially increases potential conflict but, after some point, more diversity implies inferior probabilities for potential conflict.

Results (see RQ in Table 10.2) capture how far the distribution of social aggregates in Virtual Patagonia is from the bipolar case. The idea is simple: polarization is related to the alienation that individuals and groups feel from one another, but such alienation may be fuelled by notions of within-group identity. There is intuitively a much greater risk of social tension and competition if a 5 % minority group is concentrated in one particular region of the country than if it were dispersed evenly across the country. In the Virtual Patagonia case study, demographic polarization attains higher values when the world has the more abundant resources, and when fractionalization has low values because most agents belong to group 1 or group 2. These results are different then to the expected increased territoriality as a consequence of resource scarcity and spatial concentration. From the Late Holocene onwards, the social aggregation in Patagonia was too differentiated, and their size was too reduced to allow for the emergence of social tension, segregation and hence exclusive territoriality. Part of the explanation lies in the high degree of homogenization of the founding population. When we introduce two founding populations in the simulation, for instance *mapudungun* speakers and *gününa-chon* speakers, the results of demographic polarization are completely different.

On the other hand, when social networks were high enough to integrate a big number of previously isolated agents, social tension emerged between network embedded individuals and people without any ascription. In any case, demographic polarization values were comparatively lower in Patagonia than in other parts of the world, even when the horse complex and "tehuelchization" were at their maximum. These results can be related with the low degree of between-group violence in Patagonia inferred from physical anthropology analysis. The analysis of 100 traumatic injuries in male skulls from lower valleys of Chubut and Black rivers proved showed statistically significant temporal variations in the frequency of injuries resulting from interpersonal violence in times of decreasing resources (Barrientos and Gordón 2004; Gordón 2009; Berón 2010; Flensborg 2011). The likely competition and conflict situations that could have been generated with an alleged increase in population density in some areas "do not seem to have been resolved in a violent way beyond usual levels of violence in these societies" (Barrientos and Gordón 2004, p. 64; similar results have been obtained by Flensborg 2011). The highest frequency of injuries is detected once weapons of European origin appeared in historical times, indicating the later date of inter-group violence and the relevance of exogenous factors.

There is a correlation between the low degree of territorial competition and inter-group violence in our Virtual Patagonia, as well as in the actual historical Patagonian, where human groups never configured territories with clear-cut Euclidean boundaries and explicit segregation. Our results stress the role of 'territoriality' in terms of network embeddedness (Kim 2010); a "fractal metaphor" would help us to recognize that social aggregates overlapped (Appadurai 1996, p. 46). There was no place for delimited spaces conceived in geopolitical terms, because households aggregated in groups which had no "natural" limits. This is another factor stressing the low levels of polarization in Patagonia before European colonization. Therefore, because there is reason to suppose that the way in which groups were geographically concentrated may have important ramifications for ethnic politics, including conflict and economic development, we have spatially analysed the pattern of cultural diversity emerging from economic interaction (collective hunting), technological diffusion and cultural consensus.

We have carried out different experiments assuming the same five scenarios as before, and we have calculated hierarchical clustering of identity vectors based on a standard Ward method (refer to Fig. 10.5 for a graphical view of the clusters, and Table 10.3 for a summary of the results). The number of clusters for each scenario and each experiment has been normalized using the same Clustering Cubic Criterion. The number of "groups" does not refer to emerging networks of economic interaction, but to the clustering of agents in terms of their respective identities. We want to know whether the complexity of the configured social networks has any correlation with the emergence of a global identity and the configuration of "culturally" homogenous territories.

The number of clusters, that is, the number of differentiated cultural consensuses that may emerge after 250 simulated years of communal hunting and technological borrowing is not as important as the spatialization of the differences in identity vectors. Note that such vectors represent the commonality in goals, motivations and believes. They constrain cooperation and interaction, but they are also the result of interaction networks. Therefore, it is obvious that the higher the number of

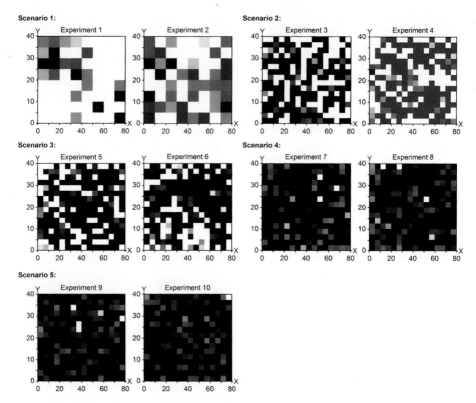

Fig. 10.5 Different scenarios of virtual Patagonia, varying the maximum resources at each patch. We have generated two different random experiments from each environmental scenario. Graphs show the spatial location of the agents after 500 time steps (ca. 250 years). *Grey* levels refer to similarity degree between agents, in such a way that each particular level reflects a similarity cluster. Clusters have been calculated using Wards method and Euclidean Distances between identity vectors (using JMP 10 Software, SAS, Inc). Table 10.3 furthermore gives the results for the above experiments in quantitative form

Table 10.3 Experimental results for the different scenarios shown in Fig. 10.5

Scenario	1		2		3		4		5	
Experiment	1	2	3	4	5	6	7	8	9	10
Max. resources	15,000	15,000	20,000	20,000	25,000	25,000	40,000	25,000	40,000	25,000
Initial agents (founding households)	300	300	300	300	300	300	300	300	300	300
Total number of agents after 250 years	41	63	246	246	323	329	956	1,018	1,162	1,184
Number of clusters	5	7	15	14	16	16	12	18	20	17

connected agents, the higher the global similarity. When resources are poor and spatially constrained (Scenario 1), population is low, and the number of networks is also very low. Consequently, there is no emergence of a common identity, but rather spatially homogenous but different identities randomly spread across the geographical area. This may have happened in Patagonia during the increased residential spatial constriction from 1,000 AD onwards. When resources become more abundant and therefore less spatially constrained (Scenarios 4 and 5) the population increases and there are more opportunities for interacting economically. As a consequence networks increase in size and in spatial extension, and a trend towards cultural homogeneity begins. That may have happened in Patagonia when new imported technology from European colonization modified drastically the chances of survival and the economic interaction among groups.

In other words, the probability of a common identity emerging within a social network decreases as the number of social networks, and the number of networked agents also decrease. The higher the number of households connected, the lesser the number of differentiated identities, and hence the more permeable seem the cultural frontiers.

To sum up, differentiated *etnia* do not emerge in 250 simulated years because of the very low temporal stability we have measured and the lack of clear segregation patterns. Preliminary results show that no aggregate has a duration of greater than a generation (50 ticks or 250 simulated years), so the fact of aggregation does not influence cultural transmission to the next generation. Identity and technological knowledge flows from parents to children, but the new generation changes and adapts its identity and knowledge constantly, according to their need for cooperation at work and the consequent flows of identity and information. Similarity clusters appear to have had more extended temporal durations than aggregates and social networks of cooperation. Evidence of exclusion and segregation are conspicuously lacking. In our case, parochialism does not emerge (Bowles and Gintis 2004; Kim 2010), because network embeddedness is in the process of being created and recreated constantly, and the average threshold of social similarity aggregates is very low in most simulated cases, indicating very high levels of tolerance to the differences of others.

Those results are what we expected. Strong ethnic differentiation based on spatial segregation should be linked to the rise of a social inequality system, and this is an aspect that we have not yet included in our model. When introducing the possibilities of "leadership" and conditioning the behavior of the next step on the behavior that a "leader" will adopt, social inertia emerges, and social aggregates increase stability and temporal duration. We think that this is what happened in Patagonia.

Historical documents from the time of European contact mention diverse forms of social and political hierarchy, notably a "chief" or "cacique" (Tomé Hernández 1587, quoted after Barros 1978). Mascardi (1963 [1670]) mentions the grouping of families around a leader with prestige. In 1784, Antonio de Viedma (cited after his *diario* published in 1836) described large groups under the leadership of a high level chief with the authority to wage war and manage most tasks that required the activity of the entire group. These large groups would subdivide into smaller groups under the leadership of their own chiefs, with a recognized independence. Each chief or leader had control of a specific territory, and the members of the group could not enter into the territory of another chief without permission. Groups generally had to pay to pass through or for the right to use the resources of another territory (Musters 1964 [1872–1873]). Other travellers mention that chiefs became poor because they distributed what they had to their followers in order to satisfy them and to be celebrated as generous (Cox 1999 [1862–1863]). Given that people could choose the leader they wanted, without this distributive behavior many chiefs would have run the risk of undermining their support.

We have interpreted nineteenth century descriptions in terms of a configuration of social order with two top hierarchical levels, usually held by men, but also by women. The highest level was more irregular in time and geographical extent, but it predominated in groups of more than 1,000 people. Immediately below, a second hierarchical level where a minor leader had some degree of authority over small groups was more usual. In some cases, especially in northern Patagonia, this second hierarchical level was subdivided into lower level hierarchical ranks. The lowest social level was that of captives reduced to slavery, whose possibilities for social mobility were extremely low. The majority of group members had no access to dominance and leadership, but they were economically and politically self-sufficient, with the possibility of freely choosing the group to which they wished to belong.

In southern and central Patagonia, the authority of chiefs was probably not very great, being limited to leading the territorial mobility of the group. Chiefs were explicitly not liberated from work (Musters 1964 [1872–1873]). However, a chief's pre-eminence was well recognized by his/her followers and neighbors during their lifetime and it was even remembered after his/her death. Only in case of war would they acquire more authority, restricting the individual freedom of group members. As a result of war, captives were integrated into indigenous family groups as servants or as a kind of slave. At the end of eighteenth century, in some parts of northern Patagonia where European contact and inter-group conflict were stronger, chiefdoms were strongly consolidated, with evident hierarchical differences and elite families with succession rights (Mandrini 1992). In the nineteenth century, but probably even before, some of the chiefdoms had an important hereditary character (Vezub 2006, 2009, 2011), although we cannot easily conclude that political power and leadership were always transmitted from father or mother to son or daughter. Some families maintained the prestige and social influence of their main members for more than two or even three generations.

The next implementation of our simulation (called PSP 1.7) will include procedures for simulating leadership and political ties to analyze the way mobility and reproduction was mediated by social decisions.

10.6 Conclusions

"Ethnicity does not explain anything, it needs to be explained" (Doornbos 1991, p. 19). The same can be said about "territoriality". This is what we have tried to explore with our computer simulation of what may have happened in Patagonian history.

Our computer model of Patagonian hunter-gatherers explores the consequences of positive interaction as an adaptive advantage. Hunter-gatherers die less and survive better when they unite to hunt together. But such economic interaction had some important effects on innovation-diffusion and on cultural diversity. Survival in our virtual world is not an adaptive process, but a probabilistic one. That is to say, households do not modify their behavior to maximize survival. However, there are higher chances for surviving when agents cooperate looking for resources and sharing labor. Given that the probabilities of cooperation are conditioned on the existence of some cultural consensus, agents should be able to adapt their identity in response to the identity of agents with whom they have successful economic interactions. Identity is then under constant negotiation, and it evolves conditioned by the number and nature of agents involved in positive interaction. Given that social aggregates constantly change their internal organization, collective identity is constantly adapting. The apparent abundance, continuity, and easy access to resources all along the humanly exploitable area would have prevented human groups from increasing their productive capacity as a consequence of competition with other groups sharing the same environment. The absence (or insignificance) among Patagonian hunters of food reserves to be used during seasons of minimal economic activity is another fact that points in this direction. The mobile nature of the dispersed population was conducive to a very low level of political elaboration and hence of social stratification.

Our simulation shows that ethnicity can be understood in terms of the tendency of people with connected (or similar) traits (including physical, cultural, and attitudinal characteristics) to interact with one another more than with people with whom they have no connection (or similar features). In addition to the principle of ethnicity choice at the level of local rules of interaction, we have also introduced the principle of social influence (i.e., the more that people interact with one another, the more similar they become) which runs at the level of communication and the formation of a socio-cognitive level. This influence process produces induced ethnicity, in which the disproportionate interaction of likes with likes may not be the result of a psychological tendency but rather the result of continuous interaction.

Preliminary results clearly suggest that it is not the geographical space, and not only the distribution of resources that explains the emergence of territoriality, social aggregation and cultural differentiation phenomena, but social interaction (cooperation) and political constraints acting on social reproduction that explains both aggregation and the constant flux of identity negotiation and rebuilding. In the model, positive interaction depends on:

- the expected benefits of identity similarity,
- the expected benefits of help from others (in terms of labor),
- the expected benefits of more efficient techniques adopted from culturally similar neighbours,
- the expected benefits of mobility (the chance of finding someone in the neighborhood able to cooperate).

Acknowledgements We would like to acknowledge to Kerstin Kowarik and to Gabriel Wurzer for their invitation to contribute to this volume.

Special thanks are due to J.M. Galán and J.I. Santos (Universidad de Burgos) and J.A. Cuesta (Universidad Carlos III-Madrid) who made an important contribution in designing parts of the original algorithm that inspired the current computer implementation. Julio Vezub and Eduardo Moreno (CENPAT. CONICET. Argentina) help us with Patagonian ethnoarchaeological data.

This research has been funded by the Spanish Ministry of Science and Innovation, through Grant No.HAR2012-31036 awarded to J.A. Barceló and Project CSD2010-00034 "Social and environmental transitions: Simulating the past to understand human behavior (SimulPast)" (CONSOLIDER-INGENIO 2010 program by Spanish Ministry of Science and Innovation, see: http://www.simulpast.es). X. Vilà acknowledges financial support from grant ECO2008-04756 (Grupo Consolidado-C) from the Spanish Ministry of Science and Innovation and from grant SGR2009-0578 of the Generalitat de Catalunya. Francisco J. Miguel also acknowledges the Spanish Ministry of Science and Innovation financial support from Grant No. CSO2012-31401 (R+D+i project)

Special thanks to two anonymous reviewers and to David Uwakwe for his comments and suggestions for improving the correctness and clarity of our text.

References

Abruzzi W (1982) Ecological theory and ethnic differentiation among human populations. Curr Anthropol 23(1):13–35

Adelaar WFH, Muysken PC (2004) The Languages of the Andes, revised edition. Cambridge University Press, Cambridge

Aguilera Faúndez O (2007) Fueguian Languages. In: Miyaoka O, Sakiyama O, Krauss ME (eds) The Vanishing Languages of the Pacific Rim. Part II: Areal Surveys. Oxford University Press, New York, pp 206–220

Alesina A, Devleeschauwer A, Easterly W, Kurlat S, Wacziarg R (2003) Fractionalization. J Econ Growth 8(2):155–194

Appadurai A (1996) Modernity at Large: Cultural Dimensions of Globalization. University of Minnesota Press, Minneapolis

Axelrod R (1997) The dissemination of culture: a model with local convergence and global polarization. J Conflict Resolut 41(2):203–226

Barberena R (2008) Arqueología y biogeografía humana en Patagonia Meridional. Publicaciones Sociedad Argentina de Antropología, Buenos Aires

Barberena R, Hajduk A, Gil AF, Neme GA, Durán V, Glascock MD, Rughini AA (2011) Obsidian in the south-central Andes: geological, geochemical, and archaeological assessment of north Patagonian sources (Argentina). Quat Int 245(1):25–36

Barceló J, Del Castillo F (2012) Why Hunter and Gatherers Did Not Die More Often? Simulating Prehistoric Decision Making. In: Proceedings of the 40th Conference on Computer Applications and Quantitative Methods in Archaeology, Southampthon

Barceló J, Del Castillo F, Mameli L, Moreno E, Videla B (2009) Where does the south begin? social variability at the southern top of the world. Arctic Anthropol 45(2):46–71

Barceló J, Del Castillo F, Mameli L, Moreno E, Sáez A (2011) Patagonia: Del presente etnográfico al pasado arqueológico. Rev Arqueología Iberoam 9(1):6–39

Barceló J, Cuesta J, Del Castillo F, Galán J, Mameli L, Miguel F, Santos J, Vila X (2013a) Simulating Prehistoric Ethnicity: The Case of Patagonian Hunter-Gatherers. In: Contreras F, Melero FJ (eds) CAA '2010: Fusion of Cultures, BAR International Series, vol S2494. ArcheoPress, Oxford, pp 137–144

Barceló J, Del Castillo F, Del Olmo R, Mameli L, Miguel F, Poza D, Vila X (2013b) Social interaction in hunter-gatherer societies: simulating the consequences of cooperation and social aggregation. Soc Sci Comput Rev. doi:10.1177/0894439313511943. Preprint, published online

Barrientos G, Gordón F (2004) Explorando la relación entre nucleamiento poblacional y violencia interpersonal durante el Holoceno tardío en el noreste de Patagonia (república argentina). Magallania 32:53–69

Barrientos G, Perez SI (2002) La dinámica del poblamiento humano del sudeste de la región pampeana. Intersecciones Antropol 3:41–54

Barrientos G, Pérez I (2004) La expansión y dispersión de poblaciones del norte de Patagonia durante el Holoceno tardío: evidencia arqueológica y modelo explicativo. In: Civalero M, Fernández P, Guráieb G (eds) Contra viento y marea: Arqueología de la Patagonia, INAPL-SAA, pp 179–196

Barrientos G, Perez SI (2005) Was there a population replacement during the Late mid-Holocene in the southeastern Pampas of Argentina? archaeological evidence and paleoecological basis. Quat Int 132(1):95–105

Barrientos G, Perez SI, Bernal V, González P, Béguelin M, Del Papa M (2008) El poblamiento holocénico del sudeste de la Región Pampeana: una discusión bioarqueológica. Apuntes del Centro de Estudios Arqueol Regionales, Universidad Nacional de Rosario, vol 2, pp 97–111

Barros JM (1978) Primer testimonio de Tomé Hernández sobre las fundaciones hispánicas del Estrecho de Magallanes. In: Anales del Instituto de la Patagonia, vol 9, Universidad de Magallanes, Punta Arenas, Chile, pp 65–76

Béguelin M (2010) Tamaño corporal y temperatura ambiental en poblaciones cazadoras recolectoras del Holoceno tardio de Pampa y Patagonia. Rev Argent Antropol Biol 12(1):27–36

Beguelin M, Barrientos G (2006) Variación morfométrica postcraneal en muestras tardías de restos humanos de patagonia: una aproximación biogeográfica. Intersecciones Antropol 7:49–62

Béguelin M, Bernal V, Del Papa M, Novellino P, Barrientos G (2006) El poblamiento humano tardío del sur de Mendoza y su relación con el norte de Patagonia: una discusión bioarqueológica. An Arqueol Etnol Vol Especial 61:6–25

Belardi J, Goñi R (2006) Representaciones rupestres y convergencia poblacional durante momentos tardíos en Santa Cruz. el caso de la meseta del Stroebel. In: Fiore D, Podestá M (eds) Tramas en la Piedra. Producción y uso del arte rupestre, Sociedad Argentina de Antropología, Buenos Aires, pp 85–94

Belardi J, Espinosa S, Carballo Marina F, Barrientos G, Goñi R, Súnico A, Bourlot T, Pallo C, Tessone A, García Guraieb S, Re A, Campan P (2010) Las cuencas de los lagos Tar y San Martín (Santa Cruz, Argentina) y la dinámica del poblamiento humano del sur de patagonia: integración de los primeros resultados. Magallania 38(2):165–188

Bellelli C, Scheinsohn V, Podestá M (2008) Arqueología de pasos cordilleranos: un caso de estudio en Patagonia Norte durante el Holoceno tardío. Bol Mus Chil Precolombino 13(2):37–55

Bernal V, Pérez I, González P (2006) Variation and causal factors of craniofacial robusticity in Patagonian hunter-gatherers from the Late Holocene. Am J Hum Biol 18:748–765

Bernal V, González P, Pérez I, Pucciarelli HM (2008) Entierros humanos del noreste de Patagonia: nuevos fechados radiocarbónicos. Magallania 36(2):175–183

Bernardi L (2003) Channels of social influence on reproduction. Popul Res Policy Rev 22(5–6):527–555

Berón M (2005) Integración de evidencias para evaluar dinámica y circulación de poblaciones en las fronteras del río Colorado. Programa y Resúmenes de las VI Jornadas de Arqueología de la Patagonia. Universidad de Magallanes, Punta Arenas

Berón M (2010) Circuitos Regionales y conflictos intergrupales prehispánicos. evidencias arqueológicas de violencia y guerra en la pampa occidental Argentina. In: Actas del XVII Congreso Nacional de Arqueología Chilena, vol 1, pp 493–503

Binford L (2001) Constructing Frames of Reference. An Analytical Method for Archaeological Theory Building Using Ethnographic and Environmental Data Sets. University of California Press, Berkeley

Bjerck H, Zangrando A (2013) Marine ventures: comparative perspectives on the dynamics of early human approaches to the seascapes of Tierra del Fuego and Norway. J I Coast Archaeol 8(1):79–90

Bodner M, Perego UA, Huber G, Fendt L, Röck AW, Zimmermann B, Olivieri A, Gómez-Carballa A, Lancioni H, Angerhofer N, Bobillo M, Corach D, Woodward S, Salas A, Achilli A, Torroni A, Bandelt H, Parson W (2012) Rapid coastal spread of first Americans: novel insights from South America's Southern Cone mitochondrial genomes. Genome Res 22(5):811–820

Bongaarts J, Watkins S (1996) Social interactions and contemporary fertility transitions. Popul Dev Rev 22(3):639–682

Borgatti S, Halgin D (2011) Mapping Culture: Freelists, Pilesorting, Triads and Consensus Analysis. In: Schensul J, LeCompte M (eds) The Ethnographer's Toolkit, vol 3. Altamira Press, Walnut Creek

Borrero L (2005) The Archaeology of Patagonian Deserts: Hunter-Gatherers in a Cold Desert. In: Veth P, Smith M, Hiscock P (eds) Desert Peoples. Archaeologucal Perspectives. Blackwell, Oxford, pp 142–158

Borrero L (2008) Early Occupations in the Southern Cone. In: The Handbook of South American Archaeology. Springer, Berlin, pp 59–77

Borrero LA (2012) The Human Colonization of the High Andes and Southern South America During Cold Pulses of the Late Pleistocene. In: Hunter-Gatherer Behaviour: Human Response During the Younger Dryas, Left Coast Press, Walnut Creek, pp 57–78

Borrero LA, Barberena R (2006) Hunter-gatherer home ranges and marine resources. Curr Anthropol 47(5):855–868

Borrero LA, Franco NV (1997) Early Patagonian hunter-gatherers: subsistence and technology. J Anthropol Res 53(2):219–239

Borrero L, Charlin J, Barberena R, Martin F, Borrazzo K, L'Heureux L (2008) Circulación humana y modos de interacción al sur del río Santa Cruz. In: Borrero L, Franco N (eds) Arqueología del extremo sur del continente americano. Resultados de nuevos proyectos, CONICET e Instituto Multidisciplinario de Historia y Ciencias Humanas, pp 155–175

Borrero LA, Barberena R, Franco NV, Charlin J, Tykot RH (2009) Isotopes and rocks: geographical organisation of southern Patagonian hunter-gatherers. Int J Osteoarchaeol 19(2):309–327

Boschín M (2002) Indigenous History of Northwest Patagonia. Regional Identities During the Seventeenth and Eighteenth Centuries. In: Briones C, Lanata JL (eds) Archaeological and Anthropological Perspectives on the Native Peoples of Pampa, Patagonia, and Tierra del Fuego to the Nineteenth Century. Bergin and Garvey Publishers, Westport, pp 7–8

Boschín MT (2001) Original Peoples. Archaeology of Northern Patagonia. In: Boschín MT, Casamiquela R (eds) Patagonia, 13,000 years of History. Leleque Museum, Emecé Editores, Buenos Aires, pp 6–4

Boschín M, Andrade A (2011) Poblamiento de Patagonia septentrional Argentina durante el Holoceno tardio: paleoambientes e imperativos sociales. Zephirus LXVIII:41–61

Boschín M, Del Castillo F (2005) El Yamnago: del registro histórico al registro arqueológico. Rev Esp Antropol Am 35:99–116

Bossert W, D'Ambrosio C, La Ferrara E (2011) A generalized index of fractionalization. Economica 78:723–750

Bowles S, Gintis H (2004) Persistent parochialism: trust and exclusion in ethnic networks. J Econ Behav Organ 55(1):1–23

Boyd R, Richerson P (2006) Culture and the Evolution of the Human Social Instincts. In: Levinson S, Enfield N (eds) Roots of Human Sociality. Berg, Oxford

Brown GK, Langer A (2010) Conceptualizing and measuring ethnicity. Oxf Dev Stud 38(4): 411–436

Brown CH, Holman EW, Wichmann S, Velupillai V (2008) Automated classification of the world's languages: a description of the method and preliminary results. STUF Lang Typology Univers 61(4):285–308

Campbell L (1997) American Indian Languages. The Historical Linguistics of Native America. Oxford Studies in Anthropological Linguistics, Oxford University Press, New York

Cardillo M (2011) Exploring latitudinal patterns of lithic technology variation in continental coastal Patagonia, Argentina. J Archaeol Sci 38(10):2675–2682

Carley K (1991) A theory of group stability. Am Sociol Rev 56(3):331–354

Carlón F (2010) Leadership and indigenous socio-political organizations in Pampas and Northern Patagonia during the 18th century: a reconstruction from inter-ethnic relationships in the Buenos Aires frontier. Rev Colomb Antropol 46(2):435–464

Casamiquela RM (1965) Rectificaciones y ratificaciones hacia una interpretación definitiva del panorama etnológico de la Patagonia y área septentrional adyacente. Instituto de Humanidades, Universidad Nacional del Sur, Argentina

Casamiquela RM (1983) Nociones de gramática del gününa küne. Centre National de la Recherche Scientifique P.U.F., Paris

Castellano C, Fortunato S, Loreto V (2009) Statistical physics of social dynamics. Rev Mod Phys 81(2):591

Cavalli-Sforza L, Menozzi P, Piazza A (1994) The History and Geography of the Human Genes. Princeton University Press, Princeton

Chakravarty SR, Maharaj B (2011) Measuring ethnic polarization. Soc Choice Welfare 37(3): 431–452

Chandra K, Wilkinson S (2008) Measuring the effect of "ethnicity". Comp Polit Stud 4/5:515–563

Charlin J, Borrero LA (2012) Rock Art, Inherited Landscapes, and Human Populations in Southern Patagonia. In: A Companion to Rock Art. Wiley, London, pp 381–397

Charlin J, González-José R (2012) Size and shape variation in Late Holocene projectile points of Southern Patagonia: a geometric morphometric study. Am Antiq 77(2):221–242

Childs H (1936) El Jimmy, Outlaw of Patagonia. JB Lippincott, Philadelphia

Claraz J (1988 [1865–1866]) Diario de viaje de exploración al Chubut. Ediciones Marymar, Buenos Aires, Marymar

Cobos V, Della Negra C, Bernal V (2012) Patrones de variación morfométrica craneofacial en poblaciones humanas del noroeste de patagonia durante el holoceno tardío. Rev Argent Antropol Biol 14(1):5–16

Cox GE (1999 [1862–1863]) Viaje en las regiones septentrionales de la Patagonia. Elefante Blanco, Buenos Aires

Currie TH, Mace R (2009) Political complexity predicts the spread of ethnolinguistic groups. Proc Natl Acad Sci USA 106:7339–7344

D'Andrade R (1987) The Development of Cognitive Anthropology. Cambridge University Press, New York

de Orbigny A (2002 [1833]) Viaje a la América Meridional. Instituto Francés de Estudios Andinos/Plural Publications, La Paz

Del Castillo F (2012) Modelando la heterogeneidad Étnica y la diversidad cultural en arqueología de cazadores recolectores patagónicos. aproximaciones desde la simulación computacional y los modelos basados en agentes. Ph.D. thesis, Autonomous University of Barcelona

Del Castillo F, Barceló J (in press) Simulando etnicidad y etnogénesis en grupos cazadores-recolectores. patagonia como caso de estudio. Libros de la Serie de Arqueología Americana, British Archaeological Report - BAR

Del Castillo F, Mameli L, Barceló J (2011) La arqueología patagónica y la reconstrucción de la historia indígena. Rev Esp Antropol Am 41(1):27–50

Del Castillo F, Barceló J, Mameli L, Miguel F, Vila X (in press) Modeling mechanisms of cultural diversity and ethnicity in hunter-gatherers. J Archeol Method Theory 13

Doornbos M (1991) Linking the Future to the Past – Ethnicity and Pluralism. In: Salih MMA, Markasis J (eds) Ethnicity and the State in Eastern Africa. Elanders Gotab, Stockholm

Dow MM, Burton ML, Reitz K, White DR (1984) Galton's problem as network autocorrelation. Am Ethnol 11(4):754–770

Duclos J, Esteban J, Ray D (2004) Polarization: concepts, measurement, estimation. Econometrica 72(6):1737–1772

Eff E (2004) Spatial and Cultural Autocorrelation in International Datasets. Technical Report 200401, Middle Tennessee State University, Department of Economics and Finance. http://ideas.repec.org/p/mts/wpaper/200401.html

Escalada F (1949) El complejo tehuelche. Estudios de etnografía patagónica. Instituto Superior de Estudios Patagónicos

Esteban J, Ray D (1994) On the measurement of polarization. Econometrica 62(4):819–851

Fabra M, DeMarchi D (2011) Geographic patterns of craniofacial variation in pre-hispanic populations from the Southern Cone of South America. Hum Biol 83(4):491–507

Fabre A (2005) Diccionario etnolingüístico y guía bibliográfica de los pueblos indígenas Sudamericanos. Self-published (online). http://www.ling.fi/Diccionario%20etnoling.htm

Favier Dubois C (2004) Fluctuaciones climáticas referibles al Período Cálido Medieval en Fuego-Patagonia. indicadores indirectos y el aporte de modelos climáticos. In: Actas de las V Jornadas de Arqueología de la Patagonia, Instituto Nacional de Antropología y Pensamiento Latinoamericano (INAPL), pp 545–556

Fearon JD (2003) Ethnic structure and cultural diversity by country. J Econ Growth 8:195–222

Fernandez Garay A (1998) El tehuelche. Una lengua en vías de extinción, Anejos de Estudios Filológicos, vol 15. Universidad Austral de Chile, Valdivia

Fernandez Garay A (2004) Diccionario tehuelche-español/ Índice español-tehuelche, CNWS Publications, vol 138/Indigenous Languages of Latin America (ILLA). Research School of Asian, African and Amerindian Studies, Universiteit Leiden, Leiden

Fernández Garay A, Hernandez GB (2004) La terminologia de parentesco entre los tehuelches ó aonek'enk de la Patagonia Argentina. Anclajes 155:121–151

Fiore D (2006) Poblamiento de imágenes: arte rupestre y colonización de la Patagonia. variabilidad y ritmo de cambio en tiempo y espacio. In: Fiori D, Podestá MM (eds) Tramas en la Piedra. Producción y usos del arte rupestre, WAC-SAA-INAPL, Buenos Aires, pp 4–2

Fiske P (1991) Structures of Social Life: The Four Elementary Forms of Human Relations. Free Press (Macmillan), New York

Fitz-Roy R (1932 [1833–1839]) Narración de los viajes de levantamiento de los buques de su Majestad Adventure y Beagle, años 1826 a 1836 (Tomos I y II). Biblioteca oficial de Marina

Flensborg G (2011) Lesiones traumáticas en cráneos del sitio Paso Alsina 1. explorando indicadores de violencia interpersonal en la transición Pampeano-Patagónica Oriental (Argentina). Intersecciones Antropol 12:45–59

Frank A (2012) Los fogones en la meseta central de Santa Cruz durante el Pleistoceno final. Magallania 40(1):145–162

García F, Moraga M, Vera S, Henríquez H, Llop E, Aspillaga E, Rothhammer F (2006) mtDNA microevolution in Southern Chile's archipelagos. Am J Phys Anthropol 129:473–481

García-Bour JA, Pérez-Pérez S, Álvarez E, Fernández A, López-Parra E, Arroyo-Pardo, Turbón D (2003) Early population differentiation in extinct aborigines from Tierra del Fuego-Patagonia: ancient mtDNA sequences and Y-chromosome STR characterization. Am J Phys Anthropol 123:361–370

Garro L (2000) Remembering what one knows and the construction of the past: a comparison of cultural consensus theory and cultural schema theory. Ethos 28(3):275–319

Giuliano P, Spilimbergo A, Tonon G (2006) Genetic, Cultural and Geographical Distances. Technical Report IZA Discussion Papers 2229, Institute for the Study of Labor (IZA). http://ideas.repec.org/p/cpr/ceprdp/5807.html

Gómez Otero J (2007) Dieta, uso del espacio y evolución en poblaciones cazadoras-recolectoras de la costa centro-septentrional de Patagonia durante el Holoceno medio y Tardío. Ph.D. thesis, Facultad de Filosofía y Letras, Universidad de Buenos Aires

Gómez Otero J, Banegas A, Goye S, Franco N (2009) Variabilidad morfológica de puntas de proyectil en la costa centro-septentrional de Patagonia Argentina: primeros estudios y primeras preguntas. In: VIII Congreso de Historia Social y Política de la Patagónia argentino-chilena. Las fuentes en la construcción de una historia patagónica, Trevelin, pp 110–118

Goñi R, Barrientos G (2004) Poblamiento tardío y movilidad en la cuenca del lago Salitroso. In: Civalero M, Fernández P, Guraieb A (eds) Contra Viento y Marea. Arqueología de Patagonia, INAPL - Sociedad Argentina de Antropología, Buenos Aires, pp 313–324

Goñi RA, Cassiodoro G, Rindel D, Bourlot T, Garcia Guraieb S, Re A, Tessone A, Aragone A, Nuevo Delaunay A, Flores J (2007) Distribución del registro arqueológico en cuencas lacustres del noroeste de la provincia de Santa Cruz. In: Proceeding of XVI Congreso Nacional de Arqueología Argentina III, Universidad Nacional de Jujuy, Argentina

Gonzalez PN, Perez SI, Bernal V (2011) Ontogenetic allometry and cranial shape diversification among human populations from South America. Anat Rec 294(11):1864–1874

Gonzalez-Jose R (2003) El poblamiento de la Patagonia. Análisis de la variación craneofacial en el contexto del poblamiento americano. Ph.D. thesis, Universitat de Barcelona

Gonzalez-Jose RG, Dahinten S, Hernández M (2001) The settlement of Patagonia: a matrix correlation study. Hum Biol 72(2):233–248

Gonzalez-José R, Bortolini M, Santos F, Bonatto S (2008) The peopling of America: craniofacial shape variation on a continental scale and its interpretation from an interdisciplinary view. Am J Phys Anthrop 137:175–187

Gordón F (2009) Atribución causal a traumas craneofaciales en muestras del Norte de Patagonia (República Argentina): una perspectiva experimental. Magallania 37(2):57–76

Guichón R (2002) Biological Anthropology in Fuego-Patagonia. In: Briones C, Lanata J (eds) Archaeological and Anthropological Perspectives on the Native Peoples of Pampa, Patagonia, and Tierra del Fuego to the Nineteenth Century. Bergin & Garvey Publishers, Westport, pp 13–30

Harrington T (1946) Contribución al estudio del indio Gününa Küne. Rev Mus Plata 2:237–275

Holman EW, Schulze C, Stauffer D, Wichmann S (2007) On the relation between structural diversity and geographical distance among languages: observations and computer simulations. Linguist Typology 11(2):393–422

Kalick S, Hamilton T (1986) The matching hypothesis reexamined. J Personal Soc Psychol 51: 73–82

Kim JW (2010) Evolution of cultural groups and persistent parochialism structure and dynamics. J Anthropol Relat Sci 4(2). http://escholarship.org/uc/item/2cr2681m

Kolo P (2012) Measuring a New Aspect of Ethnicity - The Appropriate Diversity Index. Ibero America Institute for Econ. Research (IAI) Discussion Papers 221, Ibero-America Institute for Economic Research

Lalueza C, Pérez-Pérez A, Prats E, Cornudella L, Turbón D (1997) Lack of founding Amerindian mitochondrial DNA lineages in extinct aborigines from Tierra del Fuego-Patagonia. Hum Mol Genet 1(6):41–46

Lazzari A, Lenton D (2000) Etnología y nación: facetas del concepto de Araucanización. Avá Rev Antropol 1:125–140

Lewis CM, Lizárraga B, Tito RY, López P, Iannacone G, Medina A, Martínez R, Polo S, De La Cruz A, Cáceres AM, Stone AC (2007) Mitochondrial DNA and peopling of South America. Hum Biol 79(2):159–178

L'Hereux GL (2006) El estudio arqueológico del proceso coevolutivo entre las poblaciones humanas y las poblaciones de guanaco en patagonia meridional y norte de tierra del fuego. Ph.D. thesis, Facultad de Ciencias Naturales y Museo

L'Heureux L, Franco N (2002) Ocupaciones humanas en el Área de Cabo Vírgenes (Pcia de Santa Cruz, Argentina): El sitio Cabo Vírgenes 6. Anales Del Instituto De La Patagonia (Serie Ciencias Humanas), vol 30. Univ Austral de Chile, pp 183–202

Llop E, Harb Z, Moreno R, Rothhammer R (2002) Genetic marker variation in coastal populations of Chile. Homo 53(2):170–177

Long J, Kittles R (2003) Human genetic diversity and the nonexistence of biological races. Hum Biol 75:449–471

López L, Silveira M, Stern C (2010) Uso de obsidianas entre los cazadores-recolectores en el bosque del lago Traful, Noroeste de la Patagonia, Argentina. Magallania 38(1):237–245

Luna L (2008) Estructura demográfica, estilo de vida y relaciones biológicas de cazadores recolectores en un ambiente de desierto. Sitio Chenque I (Parque Nacional Lihué Calel, provincia de la Pampa, Argentina). No. 1886 in BAR International Series, ArcheoPress, Oxford

Mancini MV, Franco NV, Brook GA (2013) Palaeoenvironment and early human occupation of southernmost South America (South Patagonia, Argentina). Quat Int 299:13–22

Mandrini RJ (1991) Procesos de especialización regional en la economía indígena pampeana (s. xviii-xix), el caso del suroeste bonaerense. Bol Am 41:113–136

Mandrini R (1992) Indios y fronteras en el área pampeana (siglos xvi-xix) balance y perspectivas. Anu IEhS 7:59–72

Martínez G, Flensborh G, Bayala P (2013) Chronology and human settlement in northeastern Patagonia (Argentina): patterns of site destruction, intensity of archaeological signal, and population dynamics. Quat Int 301:123–134

Martinic M (1995) Los Aónikenk, Historia y Cultura. Ediciones de la Universidad de Magallanes, Punta Arenas, Chile

Mascardi N (1963 [1670]) Carta y relación. In: Furlong G (ed) Nicolás Mascardi y Su Carta-Relación, Teoría

Mazia N, Chaneton EJ, Dellacanonica C, Dipaolo L, Kitzberger T (2012) Seasonal patterns of herbivory, leaf traits and productivity consumption in dry and wet Patagonian forests. Ecol Entomol 37(3):193–203

Mazzanti D (2006) La constitución de territorios sociales durante el Holoceno tardío. el caso de las sierras de tandilia, Argentina. Relac Soc Argent Antropol XXXI:277–300

Mena F (1997) Middle to Late Holocene Adaptations in Patagonia. In: McEwan C, Borrero L, Prieto A (eds) Natural History, Prehistory and Ethnography at the Uttermost End of the Earth. British Museum Press, London

Méndez CA, Stern CR, Reyes OR, Mena LF (2012) Transporte de larga distancia de obsidianas en Patagonia centro-sur durante el Holoceno temprano. Chungará (Arica) 44(3):363–375

Miotti L, Salemme M (2003) When Patagonia was colonized: people mobility at high latitudes during Pleistocene/Holocene transition. Quat Int 109:95–111

Montalvo J, Reynal-Querol M (2002) Why Ethnic Fractionalization? Polarization, Ethnic Conflict and Growth. Economics Working Papers 660, Department of Economics and Business, Universitat Pompeu Fabra. http://ideas.repec.org/p/upf/upfgen/660.html

Montalvo JG, Reynal-Querol M (2005) Ethnic diversity and economic development. J Dev Econ 76(2):293–323

Morello F, Borrero L, Massone M, Stern C, García-Herbst A, McCulloch R, Arroyo-Kalin M, Calas E, Torres J, Prieto A, Martinez I, Bahamonde G, Cardenas P (2012) Hunter-gatherers, biogeographic barriers and the development of human settlement in Tierra del Fuego. Antiquity 86:71–87

Moreno E, Izeta A (1999) Estacionalidad y subsistencia indígenas en Patagonia Central según los viajeros de los siglos xvi-xvii. In: Soplando en el viento, Universidad Nacional del Comahue, Neuquén, pp 477–490

Moreno E, Zangrando A, Tessone A, Castro A, Panarello H (2011) Isótopos estables, fauna y tecnología en el estudio de los cazadores - recolectores de la costa norte de Santa Cruz. Magallania 39(1):265–276

Müller A, Velupillai V, Wichmann S, Brown CH, Brown P, Holman EW, Bakker D, Belyaev O, Egorov D, Mailhammer R, Grant A, Yakpo K (2009) ASJP World Language Tree of Lexical Similarity: Version 2. http://email.eva.mpg.de/~wichmann/language_tree.htm

Musters G (1964 [1872–1873]) Vida entre los patagones. Un año de excursiones por tierras no frecuentadas desde el Estrecho de Magallanes hasta el Río Negro. Edciones Solar/Librería Hachette, Buenos Aires

Nacuzzi L (1998) Identidades impuestas. Tehuelches, Auca y Pampas en el Norte de la Patagonia. SAA, Buenos Aires

Nacuzzi L (2007) Los grupos nómades de la Patagonia y el Chaco en el siglo xviii: identidades, espacios, movimientos y recursos económicos ante la situación de contacto. una reflexión comparativa. Chungara Rev Antropol Chil 39(2):221–234

Nacuzzi LR (2008) Repensando y revisando el concepto de cacicazgo en las fronteras del sur de América (Pampa y Patagonia). Rev Esp Antropol Am 38(2):75–95

Neme G, Gil A, Garvey R, Llano C, Zangrando A, Franchetti F, De Francesco C, Michieli C (2011) El registro arqueológico de la Gruta del Manzano y sus implicancias para la arqueología de Nordpatagonia. Magallania 39(2):243–265

Nettle D (1999) Linguistic Diversity. Oxford University Press, Oxford

Nichols J (1997) Modeling ancient population structures and movement in linguistics. Annu Rev Anthropol 26:359–384

Nichols J (2008) Language spread rates and prehistoric American migration rates. Curr Anthropol 49(6):1109–1117

Orden M (2010) La frase verbal en gününa iajëch. un enfoque preliminar. In: Censabella M, González R (eds) Libro de Actas II de Lenguas Indígenas Americanas y II Simposio Internacional de Lingüística Amerindia, ALFAL, CONICET, Buenos Aires, cD-ROM

Orquera LA (1987) Advances in the archaeology of the Pampa and Patagonia. J World Prehistory 1(4):333–413

Orquera L, Gómez Otero J (2007) Los cazadores-recolectores de las costas de Pampa, Patagonia y Tierra del Fuego. Relac Soc Argent Antropol 32:75–99

Palermo M (2000) A través de la frontera. economía y sociedad indígenas desde el tiempo colonial hasta el siglo xix. In: Tarragó M (ed) Nueva Historia Argentina, vol 1. Editorial Sudamericana, Buenos Aires, pp 343–382

Papp C (2002) Die Tehuelche. Ein Ethnohistoricher Beitrag zu einer Jahrhundertelangen Nicht-Begegnung. Ph.D. thesis, Universität Wien

Paruelo JM, Jobbágy EG, Sala OE (1998) Biozones of Patagonia (Argentina). Ecol Austral 8(2):145–153

Perez SI (2011) Poblamiento humano, diferenciación ecológica y diversificación fenotípica en América. Runa 32(1):83–104

Pérez S, Bernal V, González P (2007) Morphological differentiation of aboriginal human population from Tierra del Fuego (Patagonia): implications for South American peopling. Am J Phys Anthropol 133(4):1067–1079

Perez S, Della Negra C, Novellino P, Gonzalez P, Bernal V, Cuneo E, Hajduk A (2009) Deformaciones artificiales del cráneo en cazadores-recolectores del Holoceno medio-tardío del Noroeste de Patagonia. Magallania 37(2):77–90

Politis GG, Prates L, Merino ML, Tognelli MF (2011) Distribution parameters of guanaco (lama guanicoe), pampas deer (Ozotoceros bezoarticus) and marsh deer (Blastocerus dichotomus) in Central Argentina: archaeological and paleoenvironmental implications. J Archaeol Sci 38(7):1405–1416

Posner DN (2004) Measuring ethnic fractionalization in Africa. Am J Polit Sci 48(4):849–863

Prates L (2008) Los antiguos habitantes del río Negro desde una visión arqueológica. Ediciones de la Sociedad Argentina de Antropología, Buenos Aires

Prates L (2009) El uso de recursos por los cazadores-recolectores posthispánicos de Patagonia continental y su importancia arqueológica. Relac Soc Argent Antropol XXXIV:201–229

Ramasco J (2007) Social inertia and diversity in collaboration networks. Eur Phys J Spec Top 143(1):47–50

Rivals F, Rindel D, Belardi JB (2013) Dietary ecology of extant guanaco (Lama guanicoe) from Southern Patagonia: seasonal leaf browsing and its archaeological implications. J Archaeol Sci 40(7):2971–2980

Romney AK (1999) Cultural consensus as a statistical model. Curr Anthropol 40:103–115
Romney AK, Weller S (1984) Predicting informant accuracy from patterns of recall among individuals. Soc Netw 4:59–77
Romney A, Kimball S, Weller J, Batchelder W (1986) Culture as consensus: a theory of culture and informant accuracy. Am Anthropol 88:313–338
Rothhammer F, Llop E (2004) Poblaciones Chilenas: Cuatro décadas de Investigaciones. Editorial Universitaria, Santiago de Chile
Schulze ED, Mooney HA, Sala OE, Jobbagy E, Buchmann N, Bauer G, Canadell J, Jackson R, Loreti J, Oesterheld M, Ehleringer J (1996) Rooting depth, water availability, and vegetation cover along an aridity gradient in Patagonia. Oecologia 108(3):503–511
Sieck WR (2010) Cultural Network Analysis: Method and Application. In: Schmorrow D, Nicholson D (eds) Advances in Cross-Cultural Decision Making. CRC Press/Taylor & Francis, Boca Raton, pp 260–269
Silveira M, López L, Pastorino G (2010) Movilidad, redes de intercambio y circulación de bienes en el sudoeste del Neuquén (Norpatagonia, Argentina): Los moluscos marinos del lago Traful. Intersecciones Antropol 11(2):227–236
Soriano A, Volkheimer W, Walter H, Box EO, Marcolin AA, Vaierini JA, Movia C, Vasina WG (1983) Deserts and Semi-Deserts of Patagonia. In: Ecosystems of the World. Temperate Deserts and Semi-Deserts. Elsevier, Amsterdam
Spegazzini C (1884) Costumbres de los patagones. An Soc Cient Argent 17:221–240
Stein J (1997) How institutions learn: a socio-cognitive perspective. J Econ Issues 31(3):729–740
Suárez J (1970) Clasificación interna de la familia lingüística chon. AILM 10:29–59
Taylor C, Hudson M (1972) The World Handbook of Political and Social Indicators, 2nd edn. Yale University Press, New Haven
Vezub J (2006) Lenguas, territorialidad y etnicidad en la correspondencia de Valentín Saygüeque hacia 1880. Intersecciones Antropol 7:28–24
Vezub JE (2009) Valentín Saygüeque y la gobernación indígena de Las Manzanas. Poder y etnicidad en la Patagonia septentrional (1860–1881). Prometeo Libros, Buenos Aires
Vezub JE (2011) Llanquitruz y la "máquina de guerra" ma puche-tehuelche. continuidades y rupturas en la geopolítica indígena patagónica (1850–1880). Antíteses 4(8):613–642
Viedma A (1836 [1784]) Diario y descripción de la Costa Meridional del Sur llamada vulgarmente Patagonia. Biblioteca Virtual Miguel de Cervantes, Buenos Aires. http://bib.cervantesvirtual.com/servlet/SirveObras/49250514294669165350046/
Viega Barros J (2005) Voces en el viento. Raíces lingüísticas de la Patagonia. Mondragón, Buenos Aires
Vigdor JL (2002) Interpreting ethnic fragmentation effects. Econ Lett 75:271–276
Villar D, Jiménez J (2003) La tempestad de la guerra: conflictos indígenas y circuitos de intercambio: elementos para una periodización (Araucanía y las Pampas, 1780–840). In: Mandrini R, Paz CD (eds) Las fronteras hispanocriollas del mundo indígena latinoamericano en los siglos XVIII-XIX:un estudio comparativo, Universidad Nacional del Comahue, Neuquén, pp 123–171
Weller SC (2007) Cultural consensus theory: applications and frequently asked questions. Field Methods 19(4):339–368
Whitmeyer J (1997) Endogamy as a basis for ethnic behavior. Sociol Theory 15(2):162–178
Wichmann S, Holman E, Stauffer D, Brown C (2008) Similarities Among Languages of the Americas: An Exploration of the WALS Evidence. http://email.eva.mpg.de/~wichmann/SantaBarbWichmannRevSubmit.pdf
Wilensky U (1999) NetLogo. Center for Connected Learning and Computer-Based Modeling, Northwestern University, Evanston. http://ccl.northwestern.edu/netlogo
Zúñiga F (2006) Mapundungun. El habla de los mapuches. Centro de Estudios Públicos, Santiago de Chile

Part IV
Summary and Outlook

Chapter 11
How Did Sugarscape Become a Whole Society Model?

André Costopoulos

11.1 Introduction

The goal of archaeological simulation is to help us understand how we became us. It helps us understand past social systems and their dynamics of change so that we can better understand who we are and where we came from. The initial strategy for doing this, as proposed by Doran (1970), involved simulating realistically complex societies in realistically complex environments. In keeping with the generalizing aims of the processual archaeology of the time, the hope was that accurate models of social systems could help us understand social dynamics in general.

This effort ran counter to a deeply rooted tradition of human exceptionalism that is closely tied to historical particularism and that characterizes anthropology and western approaches in general. It casts humans as fundamentally different from the rest of nature. Its inevitable conclusion is that the tools used to study the rest of nature, including all natural systems, are unsuitable for the study of humans. The erosion of this divide, most recently by the social statistics movement of the early nineteenth century and by Darwin's (1872) contention that natural selection can explain the origin of the human moral capacity, has not been digested to any significant extent, especially in social science (Taylor 2013). Human exceptionalism continues to underlie critiques of systemic approaches to human culture and society, although there is an emerging movement in cultural anthropology, sometimes called multispecies ethnography (Kirksey and Helmreich 2010) to situate humans in their broader natural context as social beings among others (see for example Tsing 2013; Kohn 2013).

A. Costopoulos (✉)
McGill University, Montréal, QC, Canada
e-mail: andre.costopoulos@mcgill.ca

© Springer International Publishing Switzerland 2015   259
G. Wurzer et al. (eds.), *Agent-based Modeling and Simulation in Archaeology*,
Advances in Geographic Information Science, DOI 10.1007/978-3-319-00008-4_11

This widespread assumption that methods used for the study of the rest of nature could not be used to study humans made it necessary for early practitioners of simulation in archaeology to convince their colleagues that simulation of systems could in fact help us study human societies. They had to convince them that simulation could be used to study something as complex as culture. In early applications, the scope of modeling was broad, as was the scope of implications. Entire cultures, as defined and imagined by anthropologists of the day, were modeled to learn about and explain all cultures. To increase the level of believability and strengthen the argument that simulation could be used to study real societies past and present, these were usually ethnographically or archaeologically known societies. This ancestral form of archaeological simulation can be called realist-generalist. It strove for a concrete form of realism in which elements of the simulation correspond to elements of the real historical world, as opposed to a fuzzier kind of verisimilitude in which the simulation creates a believable but hypothetical world. It led to the emergence of so-called whole society models (Aldenderfer 1998; Lake 2004).

Very quickly and quite naturally, the whole society approach gave rise to a more particularist school of simulation that sought to understand specific societies in specific historical and environmental contexts. This form of simulation is ultimately concerned with general explanation, but focuses very tightly on a well-defined archaeological case to generate its general understanding. There is emphasis on the immediate goal of understanding a context in its own terms, and usually a discussion of broader applicability or at least the potential for such. The scope of modeling is broad, as entire social systems are modeled and specified as closely as possible and in as much detail as possible, but the scope of implications tends to be narrower because the immediate concern is with understanding a particular context, as a way of eventually building greater general understanding. This is a realist-particularist form of archaeological simulation, in that it puts the particular case front and center.

A third stream of archaeological simulation retains the concern for general explanation of the ancestral form, in that it seeks generally applicable explanations of archaeological phenomena, but adopts an abstract, rather than realistic portrayal of reality. Here the emphasis is on the modeling of general processes, sometimes in no particular context at all. The particular case either becomes one of many instances of the general phenomenon, or sometimes disappears completely. Following applications in the biological and material sciences, this approach seeks to model very specific and simple processes in the pursuit of general explanation. In this form, the scope of modeling is narrow, because few processes and kinds of objects are modeled, but the scope of implications is very broad. This is an abstract-generalist form of archaeological simulation.

To at least some of its practitioners, however, this high level abstract-generalist approach has become no less a form of whole society modeling than its realist cousins. In some regions of the discipline this form has effectively become a whole society modeling of a new kind that replaces the use of context specific realism to produce historically realistic output, with process specific realism that produces verisimilitude in an attempt to increase general understanding. In this paper, I will briefly outline what I see as the main threads of the emergence of abstract-generalist simulation and its rise to the status of whole society modeling. I will try to tell in broad outlines the story of how some strands of archaeological simulation moved beyond early concerns for demonstrating that the method is useful for studying the human past. In some areas of the discipline, we see a near reversal of situation, from early days when people wondered whether computer simulations could ever be complex enough to be informative about human society, to a situation where people are wondering whether understanding human society requires anything else than very general, abstract natural-like models.

11.2   The Realist-Generalist Ancestral Form

Early simulators initially struggled against skepticism that societies are system-like. Systems are sets of elements that interact according to given rules. Computer simulations portray the operation of systems. The question of whether human societies are systems, or at least have system-like properties, has therefore been of great importance to archaeological simulation from the start. Applied to archaeology, the question acquires a chronological dimension: Is social change system-like?

The question was not new, and its modern formulations can be traced back at least to Quételet's (1835) Essaie de Physique Sociale (see also Quételet 1848). Despite some optimistic post-war statements outside the field (Stewart 1947, for example), opinion within anthropology and archaeology, especially in North America, was generally skeptical of overarching explanations of a systemic nature, as exemplified in Boas (1920). Even in the 1970s, anthropology and archaeology on both sides of the Atlantic were still reeling from the perceived excesses of classical unilinear evolutionism that had become ideologically tainted by the mid-century struggle against the extreme classical evolutionary theories of Aryan racial superiority. In that context, systems were suspect. On the other hand, institutional pressures created by expanding academic departments and the need to secure external funding were encouraging social scientists to search for general explanatory frameworks like those available to their natural science colleagues.

In keeping with Doran's (1970) manifesto for archaeological simulation, and given the debates about whether society is a system and whether computer simulation can usefully approximate it, most early simulators adopted a realistic approach and had generalist goals. They attempted to portray credibly human-like societies that were usually closely inspired from archaeologically known instances (Thomas 1973; Wobst 1974 are the classic examples). As processualists, they sought to provide insights about the mechanisms of human social organization and social change in general. There was also a marked concern for understanding how the archaeological record itself relates to our reconstructions of the past.

For the pioneers of archaeological simulation, society was unambiguously a system. The archaeological record was both the material signal of the operation of that system, as well as a system in itself, the operation of which held its own lessons for understanding the past. In his simulation of Steward's Great Basin settlement patterns, Thomas (1973, p. 174) defines a settlement-subsistence pattern as "a *system* of interrelations between loci of human occupation (sites)" (emphasis original). Wobst's (1974, p. 147) classic early effort is stimulated by the need to "integrate the particular result into a systemic whole", and to "proceed to the systemic level" of explanation of the archaeological record.

While many of their non-simulating colleagues in the heyday of processualism were receptive to the idea of society as system, they weren't always convinced that computer simulations, however elaborate and complicated they could be made with contemporary tools, could approximate the level of complexity needed to really learn about human social systems, or even that humans were capable of formulating the required hypotheses and experiments.

Even fairly friendly critics like Donn T. Bayard (1973, p. 377), who granted the basic premise that societies are systems, points out in discussing William Longacre's (1966, p. 95) suggestion that archaeologists are in the privileged position of having access to a wealth of cultural systems from the past, that "the archaeologist's 'laboratory' does not consist of cultural systems; it consists of a severely limited sample of the remains of the material manifestations of cultural systems". But Bayard goes on to suggest that even if archaeologists did have access to the entirety of the human record, it's limited size in terms of data, as opposed to even small physical systems, would be inadequate to make it amenable to study by physical science-like approaches. In other words, even granting that society is a system, and that social change is systemic, we could never prove it, much less understand it's dynamics. The amount of data available, or even potentially available, is so small that it would be impossible to derive systemic rules from the observations. Archaeology, according to this critique, was in no position to replicate the explanatory successes of the physical sciences.

Merilee Salmon (1978), an even friendlier critic, argued early on that even if human societies are systems, neither General Systems Theory nor Mathematical Systems Theory can be much use to the archaeologist, the first because it is not real theory, the second because it does not usually identify causal factors of phenomena. Interestingly, Salmon (1978, pp. 181–182) gives as an example the case of a sociology dissertation (unidentified in the text) that "explains" recent (at the time) US demographic trends by fitting the population growth curve to a model developed for fruit fly populations, pointing out that the explanation doesn't take into account factors such as "the availability of various contraceptives, the widespread publicity about the dangers of over-population, or the existence of family planning agencies". A very different reading of the same example might emphasize that it raises serious questions about whether any of those factors are relevant to the evolution of human social systems. This question is essentially the one asked my modern practitioners of the abstract-generalist approach, as I will explain below.

Hodder (1985) went farther in challenging the very idea that society is a system or even system-like. He was an especially observant and credible critic, having himself been one of the early practitioners of archaeological simulation (see Hodder 1978). His work features some of the best examples of realist-generalist simulation, including his work with Elliott and Ellman (1978) on Neolithic axe dispersal in

Britain. His main criticism of the systemic view of society is that it doesn't see humans as active participants in the construction of their society, whose goals and preferences are formed in a unique historical context. This results, in his view, in an overly deterministic model of society that ignores both culture and agency.

Certainly, Hodder accurately describes some of the processual work of the 1970s. But processual work in general, and archaeological simulation in particular, don't have to ignore the actor or the context. Taking them into account does take a different approach to modeling, one that was emerging by the mid-1980s. Full-fledged agent-based approaches, foreshadowed by cellular automata in other fields (see Gardner 1971 for a useful review of early efforts), promised to allow the archaeologist to address the main critiques raised by Hodder and others, by introducing individual choice, perspective, agency, and historical contingency to artificial societies that could help explain the human past and by implication, the human present.

The two main criticism that were levelled at the realist-generalist form of archaeological simulation were that (1) it wasn't particularly realistic and (2) that it couldn't aspire to general explanation. It was thought that the artificial social systems modeled were hopelessly inadequate to portray human societies because they lacked complexity, and especially because they lacked human-like agents. They couldn't aspire to general explanation because the social systems they sought to explain were unique and irreproducible, or at the very least, not subject to the kind of repeatable experimentation that could allow us to find out whether they were system-like.

11.3   The Realist-Particularist Approach

The increasing complexity and realism of archaeological simulation through the 1970s and 1980s, as well as the promise for more of the same, led to an approach that increased the realism of some archaeological simulations. These usually focused on very specific archaeological contexts, well bounded in time and space. The main defining characteristic of this type of simulation is its realism. It seeks to portray actual archaeologically observed systems. However, it is distributed over a wide spectrum of particularism from very weak to very strong. The strongly particularist efforts seek to understand a particular archaeological context for its own sake, with little concern for general explanation. In such strongly particularist work, there is often a sense that while two particular contexts can be interesting in their own right, one cannot help us understand another. For example, the Maya collapse is interesting and should be studied, and the Easter Island collapse is interesting and should be studied, but comparing them is not actually very informative. The weakly particularist (or more generalist) efforts tend to prioritize the study of the particular context, but their ultimate aim is to provide material for comparison, because comparison is seen as informative.

While this realist-particularist approach in archaeological simulation usually has overall generalist aims, it sometimes de-emphasizes the general implications of simulation results in favour of local understanding of a time and place. The strategy proposed is to get at the general understanding through the detailed study of a particular case, rather than using general process to explain cases. The scope of modeling here is broad, and the scope of implications is narrow.

Some early simulation efforts in anthropology pushed the realist-particularist approach to the extreme. MacLuer et al. (1971) built a microdemographic simulation of the population of four Yanomamo villages in which the simulated individuals in the starting population corresponded to actually living members of the ethnographic population. They sought "(1) to check field data for consistency and indicate areas in which more data are needed, and (2) to study demographic structure" (MacLuer et al. 1971, p. 194).

Highly realist and very particularist simulation is inevitably a large undertaking and some of the true 'mega-projects' of archaeological simulation belong to this category. They are so large in scope of modeling that they require large teams, and in some cases are well over a decade in the making. Tim Kohler's team (see Kohler and Varien 2012 for a recent statement), for example, have worked on increasingly elaborate and realistic models of social and environmental change in the American southwest. While their motivation is clearly expressed in terms of the potential for general explanation of human adaptation to environmental change, the immediate fruit of their labour is a deeper understanding of the Pueblo collapse. More significantly, they seek general explanation through detailed investigation of a particular case.

The ENKIMDU engine (Wilkinson et al. 2007) is only one aspect of a similar project has been dealing with social and environmental change in ancient Mesopotamia (Altaweel 2008; Wilkinson et al. 2007), with an emphasis on agriculture. Here again, the overall aims are generalizing, but the method is to focus on a particular case in great detail in order to get at larger questions.

11.4   The Realist-Abstract Border

A great many archaeological simulations fall within the region of realist-particularism as defined here, although it is admittedly a large region with fuzzy edges. Some of this work, while it features a particular case, doesn't push the realism quite as far the Kohler's Village project or ENKIMDU and it emphasizes general applicability to a greater degree. Lake's (2000) MAGICAL work on information sharing among early Hebridean foragers explores a particular case to learn about the social dynamics of new environment occupation and the formation of the archaeological record. MAGICAL abstracts the environment and the subsistence base of the modeled population to high degree while keeping them in a well defined archaeological context and while comparing its output to observed data. Conolly and colleagues adopt a similar strategy for their work on the spread of early agriculture in Europe (Conolly et al. 2008) and bronze age settlement patterns in the Agean (Bevan and Conolly 2011). These are just a few examples in a crowded field.

The application of computer simulation to early human evolution (at least until the Upper Paleolithic) perhaps marks the point at which the realist and the abstract regions meet. One can't be faulted for wondering whether most ABM studies that deal with pre-modern humans are about any specific case at all. The Paleolithic tends to be treated as a single, all encompassing case for which there is no comparative basis. The farther back the period with which one deals, the truer this seems to be. For example, because of the time scales involved and because of the conceptual remoteness of our ancestors, one generally models hominin dispersal (e.g. Mithen and Reed 2002; Nikitas and Nikita 2005), not the particular features of *a* dispersal for its own sake. The "case" here is a species at the global scale, not a particular population in a particular environment at a particular time. The concern, by default, is for general explanation, although it can be seen as quite particularistic if the target for explanation is the single human line of descent, and especially if hominin phenomena are treated in isolation from, or somehow differently from those related to other species.

Wobst's (1974) simulation of paleolithic populations fits into this border area. It is realistic in terms of population processes, but largely abstracts subsistence strategy, a typical target of realist simulation. It is ambiguously case-specific because it deals with human ancestors rather than *some* human ancestors. It puts general applicability front and center and treats the case in some ways as a necessary evil.

If we can agree to call Wobst's work proto-agent-based, there is another example of early simulation in archaeology that while not agent-based, brings us right to the edge of the abstract generalist approach and probably is an important precursor. Wobst used general demographic principles to model Paleolithic population, but he still intended some level of realism. Ammerman and Cavalli-Sforza (1971, 1973) were simply searching for a mathematical model that was known to be consistent with a natural process and that could be used to describe an archaeological phenomenon. They found it in Fisher's wave of advance and applied it to the spread of farming in Europe. This was part of an approach championed by Renfrew (1977) in the 1970s for the identification of families of mathematical models that could be used to describe archaeological phenomena.

Over time, simulators operating at the edges of this realist-abstract border created a firmly abstract-generalist approach. They were gradually freed from traditional archaeology's concern for the context for its own sake, and they found a like-minded audience for whom they could confidently treat society as system.

11.5 The Abstract-Generalist Approach

The abstract-generalist approach focuses on modeling high-level processes and is concerned with the specific case as an instance that illustrates the operation of the process. Whereas the realist approaches try to portray a context in some detail by modeling many processes and features, the abstract approach models few processes, sometimes a single one, and does not even necessarily tie it to a real particular case. Frederik Barth (1966) had already emphasized the central importance of generative models in the explanation of social phenomena. Generative models are those that use a set of elements and rules for interaction between them to produce a pattern or an outcome. They can be opposed to descriptive models which impose conditions. Effectively even the most generative model has descriptive elements. For example, while a model of foraging may let a settlement pattern emerge from the interaction between agents and patches, it might describe a sequence of environmental change over time by imposing a climate change curve. It would not necessarily let the climate change emerge from the interaction of climatological variables. Ultimately, no model can be completely generative.

The abstract generalist approach using generative models really started coming into its own when Epstein and Axtell (1996) provided researchers with possibly the first useable agent-based sandbox. Devoid of particular cultural, geographical, or temporal context, it allows the student of a social phenomenon to ask "how could the decentralized local interactions of heterogeneous autonomous agents generate the given regularity?" (Epstein 1999, p. 41), which Epstein calls "the generativist's question".

Since the 1970s the increasing availability of computational tools has allowed archaeologists working in that generative tradition that emphasizes process over case and general explanation over contextual understanding, to turn their models into simulations, for example in Optimal Foraging (Winterhalder 1986) or Dual Inheritance Theory (Shennan 2009), among others.

Mesoudi and Lycett (2009) is typical of this school. It features an abstract model of culture change, not tied to time and place, and explores the operation of random and frequency dependent copying in order to produce general insights that can be applied to a variety of cases. More significantly, it explicitly makes the link between human evolution and other natural processes by highlighting potential implications of their conclusions for "non-human cultural datasets" (Mesoudi and Lycett 2009, p. 47).

Hahn and Bentley (2003, p. S120) go further in arguing that society is a natural construct, although they do it through the study of a particular case. They show that changes in baby name frequency in the US are "satisfactorily explained by a simple process in which individuals randomly copy names from each other, a process that is analogous to the infinite-allele model of population genetics with random genetic drift. By its simplicity, this model provides a powerful null hypothesis for cultural change".

Most of the ABM work at the abstract end of the abstract-generalist region uses available theory to investigate the degree to which it can account for either observed or intuited social phenomena. Neo-Darwinian Theory and Economic Game Theory from of the bulk of this work. All of it assumes that society is a natural construct not that different from any other natural phenomenon. The use of theory and method derived from natural sciences, however, is not inherent to abstract-generalist approaches. There is no reason why someone might not use social theory, for example, as a framework for an abstract-generalist investigation. But such work seems to be nearly absent from the landscape, perhaps because of the interests and propensities typical of abstract-generalist workers. Vaneeckhout (2010), for example, begins a modeling effort which could easily lead to generalist-abstract simulation and which is based on Levi-Strauss' (1979) theory of house society. Whitehouse et al. (2012) build a simulation around the theory of divergent modes of religiosity to understand transmission of religious beliefs. While both efforts deal with particular case studies, their aims are clearly generalizing, and the modeling approach is abstract rather than realist.

11.6 Conclusion

In some senses, the realist position in archaeological simulation is grounded in a worry that there is something unique and difficult to account for in human social organization. In its extreme formulations, this human exceptionalist position argues that nature itself is socially constructed by humans (see Holtorff 2000–2008 for an archaeological discussion). It reflects early and ongoing concerns that computer simulation cannot adequately capture a human element necessary to the social object. It is partly a reaction to the human exceptionalist critique of the use of natural science-like method and theory for the study of humans.

The abstract position, on the other hand, expresses a conviction, at least until it can be rejected, that society is a natural construct. From that point of view, the demonstration is far from conclusive that human society is not the outcome of the natural processes that we observe at work in the evolution or the behaviour of other related species, or even of abiotic phenomena.

This is not a new debate. It was reflected in the mid-nineteenth century in the split between the Anthropological and Ethnological Societies in London. For Dunn (1861, p. 189) of the Ethnological Society, "The barrier is indeed impassable which separates man from the Chimpanzee". For Wake (1872, p. 83), the Australian Aborigines can teach us about a time when, quoting Darwin, he thinks that our ancestors had "only doubtfully attained the rank of manhood".

It is also reflected in the great controversies of the mid-twentieth century between, for example, Leslie White and the Boasians. For one side, cultural evolution could be described, if not quite explained, by energy equations. For the Boasians, humans required a more subtle treatment. It exists today in archaeological simulation in the form of the bifurcation between a realist school that continues to engage dirt-based archaeologists, and an abstract school that is increasingly speaking to more receptive audiences outside traditional archaeology, including physics, economics, evolutionary biology, and evolutionary psychology.

If the abstract position in archaeological simulation argues that human societies can be studied using the tools we have reserved for the study of the natural world, the emerging multispecies ethnography work discussed in the introduction seems to make the diametrically opposite claim that the natural world can profitably be studied using the tools we have so far reserved for documenting and understanding human societies. Does this herald a convergence of views between hard-nosed, simulation wielding evolutionary archaeologists and their free-spirited deep ethnography conducting anthropological cousins? This is an interesting question. It certainly opens up a new avenue for dialogue where such avenues have had a habit of closing over time.

References

Aldenderfer MS (1998) Quantitative methods in archaeology: a review of recent trends and developments. J Archaeol Res 6:91–120
Altaweel M (2008) Investigating agricultural sustainability and strategy in Northern Mesopotamia: results produced using a socio-ecological modeling approach. J Archaeol Sci 35:821–835
Ammerman AJ, Cavalli-Sforza LL (1971) Measuring the rate of spread of early farming in Europe. Man New Ser 6(4):674–688
Ammerman AJ, Cavalli-Sforza LL (1973) A Population Model for the Diffusion of Early Farming in Europe. In: Renfrew C (ed) The Explanation of Culture Change: Models in Prehistory. Duckworth, London, pp 343–358
Barth F (1966) Models of Social Organization. Royal Anthropological Institute of Great Britain and Ireland
Bayard D (1973) Science, theory, reality in the 'new archaeology'. Am Antiquity 34:376–384
Bevan A, Conolly J (2011) Terraced fields and Mediterranean landscape structure: an analytical case study from Antikythera, Greece. Ecol Model 222:1303–1314
Boas F (1920) The methods of ethnology. Am Anthropol 22:311–321
Conolly J, Colledge S, Shennan S (2008) Founder effect, drift, and adaptive change in domestic crop use in early neolithic Europe. J Archaeol Sci 35:2797–2804
Darwin C (1872) The Descent of Man. Project Guntenberg. http://www.gutenberg.org/cache/epub/2300/pg2300.html. e-Book
Doran JE (1970) Systems theory, computer simulations, and archaeology. World Archaeol 1:289–298
Dunn R (1861) On the physiological and psychological evidence in support of the unity of the human species. Trans Ethnol Soc Lond 1:186–202
Elliott K, Ellman D (1978) The Simulation of Neolithic Axe Dispersal in Britain. In: Hodder I (ed) Simulation Studies in Archaeology. Cambridge University Press, Cambridge
Epstein JM (1999) Agent-based computational models and generative social science. Complexity 4(5):41–60
Epstein JM, Axtell R (1996) Growing Artificial Societies: Social Science from the Bottom Up. Brookings Press/MIT Press, Washington/Cambridge/London
Gardner M (1971) On cellular automata, self-reproduction, the Garden of Eden and the game of life. Sci Am 224:112–117
Hahn M, Bentley R (2003) Drift as a mechanism for cultural change: an example from baby names. Proc R Soc B 270:S120–S123
Hodder I (ed) (1978) Simulation Studies in Archaeology. Cambridge University Press, Cambridge
Hodder I (1985) Post-Processual Archaeology. In: Schiffer M (ed) Advances in Archaeological Method and Theory, vol 8. Academic, New York
Holtorff C (2000–2008) Radical Constructivism. Electronic Monograph. https://tspace.library.utoronto.ca/citd/holtorf/3.8.html
Kirksey S, Helmreich S (2010) The emergence of multispecies ethnography. Cult Anthropol 25:545–576
Kohler TA, Varien MD (2012) Emergence and Collapse of Early Villages: Models of Central Mesa Verde Archaeology. University of California Press, Berkeley

Kohn E (2013) How Forests Think: Toward an Anthropology Beyond the Human. University of California Press, Oakland

Lake M (2014) Trends in archaeological simulation. J Archaeol Method Theory: 258–287

Lake MW (2000) MAGICAL Computer Simulation of Mesolithic Foraging. In: Kohler TA, Gumerman GJ (eds) Dynamics in Human and Primate Societies: Agent-Based Modelling of Social and Spatial Processes, Oxford University Press, New York, pp 107–143

Levi-Strauss C (1979) The Way of the Masks. Douglas and McIntyre, Toronto

Longacre W (1966) Changing patterns of social integration: a prehistoric example from the American Southwest. Am Anthropol 68:94–102

MacLuer J, Neel J, Chagnon N (1971) Demographic structure of a primitive population: a simulation. Am J Phys Anthropol 35:193–208

Mesoudi A, Lycett S (2009) Random copying, frequency-dependent copying and culture change. Evol Hum Behav 30:41–48

Mithen S, Reed M (2002) Stepping out: a computer simulation of Hominid dispersal from Africa. J Hum Evol 43:433–462

Nikitas P, Nikita E (2005) A study of hominin dispersal out of Africa using computer simulations. J Hum Evol 49:602–617

Quételet A (1835) Sur l'homme et le développement de ses facultés, ou essai de physique sociale. Bachelier Imprimeur-Libraire, Paris. https://ia600301.us.archive.org/19/items/surlhommeetled00quet/surlhommeetled00quet.pdf

Quételet A (1848) Du système social et des lois qui le régissent. Guillaumin et cie. https://ia600501.us.archive.org/19/items/dusystmesocil00quet/dusystmesocil00quet.pdf

Renfrew C (1977) Alternative Models for Exchange and Spatial Distribution. In: Earle T, Ericson J (eds) Exchange Systems in Prehistory. Academic, New York, pp 71–90

Salmon M (1978) What can systems theory do for archaeology? Am Antiquity 43:174–183

Shennan S (2009) Pattern and Process in Cultural Evolution. University of California Press, Oakland

Stewart J (1947) Suggested principles of social physics. Science 106:35

Taylor H (2013) Connecting interdisciplinary dots: Songbirds, 'white rats' and human exceptionalism. Soc Sci Inf 52:287–306

Thomas D (1973) An empirical test for Steward's model of great basin settlement patterns. Am Antiquity 38:155–176

Tsing A (2013) More Than Human Sociality. In: Hastrup K (ed) Anthropology and Nature. Routledge, London, pp 27–42

Vaneeckhout S (2010) House societies among coastal hunther-gatherers: a case study of Stone Age ostrobothnia. Nor Archaeol Rev 43:12–25

Wake C (1872) The mental characteristics of primitive man, as exemplified by the Australian Aborigines. J Anthropol Inst Great Britain Ireland 1:74–84

Whitehouse H, Kahn K, Hochberg M, Bryson J (2012) The role for simulations in theory construction for the social sciences: case studies concerning divergent modes of religiosity. Religion Brain Behav 2:182–224

Wilkinson T, Christiansen J, Ur J, Widell M, Altaweel M (2007) Urbanization within a dynamic environment: modeling Bronze Age communities in Upper Mesopotamia. Am Anthropol 109:52–68

Winterhalder B (1986) Optimal foraging: simulation studies of diet choice in a stochastic environment. J Ethnobiol 6:205–223

Wobst H (1974) Boundary conditions for paleolithic social systems: a simulation approach. Am Antiquity 39:147–178

Printed by Printforce, the Netherlands